01-HHE-1

APPLICATION OF DNA MICROARRAY TECHNOLOGY FOR WASTEWATER ANALYSIS

by:

Frédéric Berthiaume
Roland Brousseau
Karine Lemarchand
Luke Masson
Christine Maynard
Biotechnology Research Institute, Montreal, Québec, Canada

Pierre Payment
INRS — Institut Armand-Frappier, Laval, Québec, Canada

Paul Bayardelle
Centre Hospitalier Universitaire de Montréal, Québec, Canada

Josée Harel
Faculté de Médecine Vétérinaire, Université de Montréal, Québec, Canada

2004

WERF

The Water Environment Research Foundation, a not-for-profit organization, funds and manages water quality research for its subscribers through a diverse public-private partnership between municipal utilities, corporations, academia, industry, and the federal government. WERF subscribers include municipal and regional water and wastewater utilities, industrial corporations, environmental engineering firms, and others that share a commitment to cost-effective water quality solutions. WERF is dedicated to advancing science and technology addressing water quality issues as they impact water resources, the atmosphere, the lands, and quality of life.

For more information, contact:
Water Environment Research Foundation
635 Slaters Lane, Suite 300
Alexandria, VA 22314-1177
Tel: (703) 684-2470
Fax: (703) 299-0742
www.werf.org
werf@werf.org

This report was co-published by the following organization. For nonsubscriber sales information, contact:

IWA Publishing
Alliance House, 12 Caxton Street
London SW1H 0QS, United Kingdom
Tel: +44 (0) 20 7654 5500
Fax: +44 (0) 20 7654 5555
www.iwapublishing.com
publications@iwap.co.uk

Library of Congress Catalog Card Number: 2004106881

IWAP ISBN: 1-84339-727-7

Biotechnology Research Institute, National Research Council of Canada

The research on which this report is based was funded, in part, by the United States Environmental Protection Agency through Cooperative Agreement No. CR-827345-01 with the Water Environment Research Foundation (WERF). Unless an EPA logo appears on the cover, this report is a publication of WERF, not EPA. Funds awarded under the Cooperative Agreement cited above were not used for editorial services, reproduction, printing, or distribution.

ACKNOWLEDGMENTS

Report Preparation

Principal Investigator:

Roland Brousseau, Ph.D.
Biotechnology Research Institute, Montréal, Québec, Canada

Project Team:

Frédéric Berthiaume, M.Sc.
Karine Lemarchand, Ph.D.
Luke Masson, Ph.D.
Christine Maynard, M.Sc.
Biotechnology Research Institute, Montréal, Québec, Canada

Paul Bayardelle, M.D.
Centre Hospitalier Universitaire de Montréal,Montréal, Québec, Canada

Josée Harel, Ph.D.
Faculté de Médecine Vétérinaire, Université de Montréal, Québec, Canada

Pierre Payment, Ph.D.
INRS — Institut Armand-Frappier, Laval, Québec, Canada

Project Subcommittee

Linda Chrisey, Ph.D.
Naval Research Laboratory

James Fleming, Ph.D.
Center for Environmental Biotechnology, University of Tennessee

Arthur Meyers, Ph.D.
Eastman Chemical Company

Margie Nellor, R.C.
Los Angeles County Sanitation Districts

Annette Prescott, Ph.D.
Thames Water Utility, UK

Frank Schaefer, Ph.D.
U.S. Environmental Protection Agency

Water Environment Research Foundation Staff:

Director of Research: Daniel M. Woltering, Ph.D.
Project Manager: Lola Olabode, M.P.H.

ABSTRACT AND BENEFITS

Abstract:

Deoxyribonucleic acid (DNA) microarrays are widely used for differential expression studies and for detection of virulence genes in pure bacterial cultures. Their use in complex microbial samples, such as soil and wastewater, has been relatively less studied. This report presents the results of a systematic effort to apply DNA microarrays to pathogen detection and to bacterial source tracking (BST) in wastewater. Parameters such as the method of DNA extraction from the samples, the type of immobilized probe (whether polymerase chain reaction [PCR] amplicons or oligonucleotides), the length and method of immobilization of oligonucleotides, the method of DNA labelling, the combination of PCR amplification with microarray hybridization and the choice of PCR targets have been optimized. Results indicate that the combination of PCR followed by microarray hybridization can detect pathogens in wastewater samples down to a 0.1% detection limit. The use of microarrays for bacterial source tracking gave promising results on human samples, however the probes used in this study only provided signals for general indicators of fecal contamination when used on samples of animal origin. The overall conclusion is that microarray technology has not yet reached the stage of routine use for microbiological analysis of wastewater.

Benefits:

- Demonstrates that hybridization of total genomic DNA on microarrays has a high detection limit, of the order of 10^7 genomes;
- Demonstrates that the use of long oligonucleotides or PCR amplicons from 16S rDNA, or *cpn60* probes has insufficient specificity to differentiate several important pathogens, especially within the Enterobacteriaceae family;
- Demonstrates that the use of short oligonucleotide immobilized probes coupled with PCR amplification of conserved genes, such as 16S rDNA, *cpn60,* or *wecE,* can detect pathogens in wastewater down to a 0.1% (DNA weight/weight) concentration; and
- Demonstrates the potential of DNA microarrays in BST, even if further research work remains necessary to achieve this goal.

Keywords: Pathogen detection, bacterial source tracking, PCR, microarrays, 16S rDNA, *cpn60*, *wecE*

TABLE OF CONTENTS

LIST OF TABLES

Please Note: The Tables and Figures found in Chapters 4.0 and 5.0 can be viewed in color in the Product Catalog section of the WERF Website at www.werf.org.

LIST OF FIGURES

Please Note: The Tables and Figures found in Chapters 4.0 and 5.0 can be viewed in color in the Product Catalog section of the WERF Website at www.werf.org.

LIST OF ACRONYMS

A	Adenosine
bp	Base pairs
BST	Bacterial source tracking
C	Cytosine
CFU	Colony-forming units
ddH_20	Double-distilled water
DMSO	Dimethylsulfoxide
DNA	Deoxyribonucleic acid
ECA	Enterobacterial common antigen
FC:FS	Ratio of fecal coliforms to fecal streptococci
G	Guanine
GuSCN-DE	Guanidine thiocyanate-diatomaceous earth extraction
I	Inosine
ICC	Integrated cell culture
PCI	Phenol/chloroform/isoamyl alcohol
PCR	Polymerase chain reaction
rDNA	Ribosomal DNA
RNA	Ribonucleic acid
rRNA	Ribosomal RNA
S	Svedberg unit, equal to 10^{-13} second, used for expressing sedimentation coefficients of macromolecules.
SDS	Sodium dodecyl sulfate
T	Thymidine
Tm	Melting temperature
TPYB	Trypticase phytone yeast broth
tRNA	Transfer RNA
WHO	World Health Organization

EXECUTIVE SUMMARY

This project explored the utility of microarrays in wastewater analysis. An extensive literature review was first performed, which indicated that the monitoring of microorganism viability, although initially proposed, was not likely to be achieved via microarray analysis. The project team, therefore, focused on pathogen detection and bacterial source tracking in wastewater. The literature review also indicated the need for optimization of a deoxyribonucleic acid (DNA) extraction method prior to microarray analysis. Several DNA extraction protocols were compared and a standardized protocol was developed and tested.

The work on pathogen detection addressed the issues of sensitivity and specificity on various types of microarrays, including those made up of immobilized, double-stranded amplicon probes and of oligonucleotide probes of various lengths. The main findings were:

1. The use of genomic DNA is not appropriate due to the lack of sensitivity of the microarrays. The most sensitive microarrays (amplicon probes) still require 2×10^7 genomes to provide a clear signal.

2. In view of the preceding point, it is imperative that the microarray be preceded by a polymerase chain reaction (PCR) amplification step. This adds to complexity and cost, and also introduces the problem of PCR bias.

3. The use of PCR prior to microarray hybridization requires the use of short, specific oligonucleotides (22 to 26 bases long) when conserved genes such as 16S rDNA are the amplification targets. Use of longer oligonucleotides (in the 50 to 70 base range) creates cross-hybridization problems without significant compensating advantages.

4. To detect several pathogens without having to run one PCR reaction for each target pathogen, conserved genes that can be amplified with universal primers are essential. The genes investigated were 16S rDNA, *cpn60,* and, for the Enterobacteriaceae, *wecE.* Each of these genes was found to have advantages and disadvantages :

    a. 16S is broadly represented in the literature and offers targets for any desired microbial pathogen. Its highly conserved nature, however, makes its difficult to distinguish among closely related species.
    b. *cpn60* is more diverse and, therefore, more specific than 16S. Nevertheless, its universal primers are more degenerate than those for 16S, and this attribute may account for its lower sensitivity on pure culture, with a detection limit of 1000 genomes versus 100 genomes for 16S and *wecE*. Its database, however, is more limited than those for 16S rDNA.
    c. *wecE* gave superior results both for specificity and sensitivity in spiked wastewater samples. It can only be used in Enterobacteriaceae, however.

5. The detection limit for the PCR+ microarray combination is 100 genomes in pure culture (16S and *wecE*).

6. The detection limit for the PCR+ microarray combination in mixed samples including spiked wastewater DNA samples is 0.1% for *Salmonella*.

The main conclusions from the microarray work on bacterial source tracking were:

- Specific signals for human fecal indicators have been obtained from wastewater samples from human origin.
- Manure and manure runoff samples gave signal only for general fecal indicators. This finding may indicate that the animal fecal indicators represented on the microarray are either inappropriate for these samples or are present in too low a concentration (perhaps below 0.1%, based on our pathogen detection thresholds).

With current procedures, the timeframe for microarray analysis of a wastewater sample is of the order of two and a half working days. The cost of the supplies needed is of the order of $70 per experiment, providing the ability to detect simultaneously several pathogens down to a level of the order of 0.1%. Reproducibility of the protocols is fairly good, although some problems remain in the areas of reproducibility of the PCR amplification protocols and in fluorescent background control. The PCR+ microarray combination appears destined to remain primarily a research tool at the present time inasmuch as its cost, time requirements, and technical complexity seem to preclude its routine use.

CHAPTER 1.0

Literature Review on Pollution Indicators

Water quality assessment and detection of fecal indicator and pathogenic microorganisms constitute important challenges for public health and are of great significance for several public health issues:

- Detection and assessment of the degree of fecal pollution in marine water and freshwater;
- Performance assessment of the various stages of water and wastewater treatment;
- Prevention of microbial pollution of water supplies; and
- Demonstration by regular bacteriological surveillance that the high quality of water is maintained throughout the distribution system. In fact, during distribution significant physical, chemical, and microbiological changes occur. These include microbial regrowths, microbiologically induced chemical changes, and contamination.

In addition to public health, efficient water quality monitoring is a great challenge in such areas as ecotoxicology and, more recently, bioterrorism prevention. In densely populated areas, much of the contamination results from overflowing sanitary sewer systems, wastewater treatment facilities, leachate from septic tanks, and fecal matter associated with storm runoff from areas with high densities of wildlife, pets, or livestock. Nevertheless, detection, isolation, and identification of specific waterborne pathogens in natural samples are difficult, time-consuming, and hugely expensive if attempted on a regular basis. To avoid public health risks, indicator organisms are commonly used to monitor the pollution level and to predict the potential presence of pathogenic microorganisms in water samples.

1.1 Water Quality Indicators: Potential and Weaknesses

Indicator bacteria are usually harmless, abundant and originate from the same potential sources as the pathogens. Although the presence of an indicator does not prove the presence of pathogenic organisms in water, monitoring for indicator organisms is less expensive and easier than monitoring for pathogenic organisms and provides a useful indication of the microbial water quality. The World Health Organization (WHO) defines three indicator types (*1*):

- General microbial indicators demonstrating the efficacy of a process (e.g., total coliforms for assessment of chlorine disinfection);
- Fecal indicators indicating the fecal contamination (e.g., thermotolerant coliforms); and
- Index and model organisms indicating both the potential presence of a pathogen and its behaviour in the system (e.g., *Escherichia coli* as an index for *Salmonella* and *Shigella,* F-RNA (ribonucleic acid) coliphages as models for human enteric viruses).

1.1.1 General Fecal Pollution Indicators

- *Total coliform* bacteria are Gram-negative, non-spore-forming, facultative anaerobic, rod-shaped bacteria that ferment lactose to acid and gas within 24–48 h at 36 ±1°C. They belong to the family Enterobacteriaceae, which includes well-known genera such as *Escherichia,*

Klebsiella, *Enterobacter,* and *Citrobacter* (*2*) (*3*). These bacteria do not always indicate fecal pollution as many members of the group are natural inhabitants of many soils and water environments not impacted by fecal pollution.

- *Fecal coliform bacteria* (thermotolerant coliforms) are a subgroup of coliform bacteria that are able to grow at an elevated temperature (44.5 ± 0.2°C) (*2*). The presence of fecal coliforms in water indicates that fecal contamination of the water by a warm-blooded animal has occurred. *Escherichia coli* (*E. coli*) is a rod-shaped bacterium commonly found in the gastrointestinal tract and feces of warm-blooded animals. It is the dominant member of the fecal coliform group (approximately 90% of all coliforms in fecal matter are *E. coli* cells) and is distinguished by the absence of urease and the presence of β-glucuronidase in 90% of the cells (*1*). This bacterium provides direct evidence of fecal contamination from warm-blooded animals.
- *Enterococci* are a subgroup of the more general group of fecal streptococci bacteria that can be differentiated by their ability to grow at 10° and 45°C, at pH 9.6, and in a medium containing 6.5% NaCl (*4*). They are Gram-positive, catalase-negative cocci that cleave esculin and are not inhibited by bile salts. They belong to the genera *Enterococcus* and *Streptococcus* possessing the Lancefield group D antigen (*2*). These bacteria are common inhabitants of the human intestinal tract and have been found to be more resistant to stress than fecal coliforms (*5*). In addition, their survival rate through wastewater processes is similar to those of a range of potential waterborne pathogens.
- *Clostridium perfringens* is a strictly anaerobic, spore-forming, rod-shaped bacterium that is consistently found at high concentrations in human and animal feces as well as sewage (*2*). These bacteria are useful as a water-quality indicator because they form spores that are more resistant to disinfection and environmental stresses than are other indicator bacteria (*6-9*). *C. perfringens* may be used as a surrogate for stress-resistant organisms and was identified as the only indicator significantly correlated to the presence of the pathogenic protozoan *Giardia lamblia* and the opportunistic bacterial genus *Aeromonas* (*6*). The stability of the spores in natural environments can indicate remote pollution events (*10*). As they are obligate anaerobic bacteria, anaerobic conditions are required for their detection; most bacteriology laboratories should be able to perform the assay.

1.1.2 Usefulness of Indicators in Bacterial Source Tracking (BST)

The main origin of human pathogenic microorganisms is fecal matter from human or animal sources. It is generally thought that human sources pose a more significant risk as they may harbor pathogens found only in humans or that are better adapted to invasion of the human body. However, none of the most commonly used indicators are strictly related to human fecal waste. As the differentiation of human and animal fecal pollution is a great challenge for public health, targeting or combining different indicators can be helpful to determine what kind of pollution occurs. Various approaches have been proposed based on (1) speciation; (2) specific biochemical reactions; (3) profiles and ratios such as fecal coliforms vs. enterococci ratio; and, more recently, (4) DNA profiles and genotyping (*11*).

Unfortunately, none of these approaches is perfect. Many microorganisms have multiple hosts and carry out similar biochemical reactions. Moreover, horizontal gene transfers can result in variable genotypes, rendering their analysis more difficult. Nevertheless, a wide range of microorganisms has been proposed in the literature to solve this problem. These organisms include fecal streptococci, bifidobacteria, *Bacteroides* sp. (and, more specifically, *Bacteroides fragilis* phages), F-RNA phage serogroups, and *Rhodococcus coprophilus*.

Fecal streptococcal concentrations have been intensively used to monitor fecal source pollution because they are lower than fecal coliform concentrations in human sewage (10 to 100 times fewer) but higher in animal sewage. Three different approaches have been proposed to use these organisms in BST (*11*). The first is based on species identification because several authors have demonstrated that the ratio of enterococci to non-enterococci species is different among vertebrates showing a dominance of enterocci species in humans and non-enterococci species in animals (*12*). The second approach is well known and based on the ratio between fecal coliforms and fecal streptococci counts (FC:FS). As this ratio has been reported as >4 in human feces and <0.7 in animal feces (*2*), it seems possible to discriminate between a human or animal pollution source. Nevertheless, the main limitation of the usefulness of this ratio is the differential survival of the two microorganisms in receiving waters and the modification of the ratio with the residence time in water. The last approach is the use of the ratio shift, which is directly linked to the different survival rate of streptococci and coliforms. For a human source, the initial ratio decreases with time, whereas for an animal source, the initial ratio actually rises upon storage. Caution is highly recommended when using FC:FS ratio in natural waters (*13, 14*).

Bifidobacteria are one of the dominant inhabitants of human gut. Their abundance in human feces (i.e., 10^{10} CFU per gram) is generally 10 to 100 times greater than fecal coliform abundance (*15*). Nevertheless, it is difficult to use this indicator alone due to the methodological problems encountered for the detection and quantification of bifidobacteria in animals. Moreover, bifidobacteria survival capacities in water are the subject of much debate. Some authors suggest no differences exist between bifidobacteria and fecal coliforms, butr others demonstrate a weaker survival rate in water than coliforms (*15, 16*). Phenotypic analyses have demonstrated the presence of seven main groups of bifidobacteria: three originate from human feces, three from animal feces, and a specific one was found in sewage (*17*). Among them, human feces strains are both mannitol- and sorbitol-fermentative and are unable to grow at 45°C on trypticase phytone yeast broth (TPYB) (*18*).

Bacteroides sp. constitutes the second most dominant bacterial group in human gut. They are 100-fold more abundant than *E. coli* cells in human feces, but few cells are recovered from animal feces (*2, 19*). Although some specific species can be used as bacterial source indicators (*20*), the detection, isolation, and identification of these species require cumbersome and time-consuming methods. Therefore, the usefulness of *Bacteroides* sp. as fecal source indicators depends on using DNA-based techniques rather than traditional culture-based methods. One limitation to the use of these bacteria in water is related to their rapid inactivation compared to *E. coli* cells (*21*). Another way to use *Bacteroides* sp. as indicators is to target phages of the human-specific strains *Bacteroides fragilis* HSP40 and RYC4023 because they have a similar survival rate to well-known viruses (e.g., coliphage F2, poliovirus) in both freshwater and seawater and survive longer than the F-RNA phages (*2, 22, 23*). They are considered to be specific indicators of human pollution.

F-RNA phage serogroups are natural inhabitants in animal feces, found in significant concentrations in human wastewater systems; paradoxically, however, they are rarely found in human feces (*24, 25*). Serogroups II and III are mainly isolated from human feces and sewage, whereas serogroups I and IV are usually found in animal feces. Nevertheless, the question of the usefulness of these phages as fecal pollution source indicators is the subject of debate. Some studies have demonstrated that the association of specific serogroups with human and animal feces may not be as absolute as previously thought, and methods relying on the analyses of serogroups in natural systems are laborious and time-consuming (*26-28*).

Rhodococcus coprophilus is a natural inhabitant of herbivores but absent from human feces. It can be considered as a specific indicator of animal fecal pollution. Moreover, because the *R. coprophilus* survival rate in water and sediments is higher than the survival rate of *E. coli* and fecal streptococci, the ratio between *R. coprophilus* and *E. coli* increases as one moves farther downstream of human settlements (*2, 29*).

1.1.3 Conclusions about Fecal Indicators

The groups described above are generally recognized as the most commonly used fecal indicators. In addition, several specific bacterial fecal indicators can be used to differentiate human from animal fecal pollution and can, therefore, be used for bacterial source tracking in the field. Nevertheless, a lack of correlation between fecal indicators and pathogens is often observed in water systems; this can be explained by differences in survival rate, transport, and environmental factors (*5, 6, 30-33*). Also, human viral and protozoan pathogens are sometimes more persistent and not inactivated as efficiently by wastewater treatment processes as the fecal bacterial indicator groups (*7, 30*). Therefore, monitoring for fecal indicators gives a good indication of pollution levels but may not be sufficient means to predict the abundance of specific pathogens in contaminated waters.

1.2 Waterborne Pathogens

Waterborne pathogens are disease-causing bacteria, viruses, and protozoans that can be transmitted to people via the consumption of contaminated waters. Various pathogenic microorganisms can be discharged in aquatic environment through human and animal fecal contamination. Most microbial waterborne pathogens involved in human pathogenesis originate from the enteric tract of humans and animals and generate infections either by ingestion, direct skin/mucus membrane contact, or inhalation of contaminated water. The presence and concentration of these organisms in water and their public health relevance depend on 1) the amount of infection in the community; 2) the opportunity for feces to penetrate the water supply; and 3) the survival of the pathogenic enteric microorganisms in water systems.

In recent years, several pathogenic microorganisms have arisen as problems in freshwater as well as in drinking water sources, production, and distribution. Because of the public health risks associated with the presence of pathogens in water, numerous data have been made available, and it is possible to distinguish fecal pathogens from environmental pathogens. Extensive information can be found in several reviews (*34-37*).

1.3 Molecular Methods Used for Fecal Pollution Monitoring

The primary goal of water monitoring is to avoid the need for fecal indicators by using specific methods that allow direct detection of pathogens. Presently, conventional methods required for the identification of pathogens in water are based on morphological, biochemical, and serological tests. Most of these methods are laborious and require several days to be completed.

In the last two decades, scientists have developed nucleic-acid-based methods to detect indicators as well as pathogens without the growth of targeted microorganisms. These methods are much faster than culture-based methods and are able to detect targeted microorganisms with a high degree of sensitivity and specificity (*38, 39*). Identification systems are predominantly based on molecular hybridization properties involving the recognition of a targeted sequence by a nucleic acid probe. The probe can be single-stranded (oligonucleotide probes) or double-stranded (DNA probes). Hybridizations can occur between a DNA probe and specific DNA sequences (DNA-DNA hybridization) or between a DNA probe and rRNA or tRNA sequences (DNA-RNA hybridization) to assess physiological or taxonomic identification.

Extraction of nucleic acids from environmental samples is the basis of all molecular methods. Total DNA can be used to monitor the presence or absence of specific microorganisms in a complex bacterial community. The efficiency of the extraction determines the quantity, quality, and diversity of the extracted DNA and RNA. DNA can be extracted by direct lysis procedures, which disrupt the bacterial cells directly in the environmental samples, or by indirect lysis procedures, which require an initial cell extraction step and may be less efficient than direct lysis treatment. Due to the low abundance of pathogens in water, compared to the concentrations of autochthonous microorganisms, molecular analyses generally require a specific amplification of targeted sequences.

1.3.1 Polymerase Chain Reaction (PCR)

PCR amplification is an exponential amplification of any target DNA sequence that significantly increases the probability of detecting a microorganism present in low numbers. The most frequently used methods to assess the microbial diversity of water are based on PCR amplification and direct cloning or sequencing of specific targets within the microbial genomes. Unfortunately, precise enumeration of target cells is difficult, and such analyses are time consuming and laborious.

1.3.1.1 DNA Analyses

PCR is widely used for the detection of specific microorganisms. Because PCR analyses are based on the detection of intact nucleic acids rather than intact viable cells, positive PCR amplifications can arise from dead or noninfectious cells because DNA is stable enough to be amplified many days after cell viability has been lost (*40, 41*). In bacterial detection and identification, the most frequently used target is the 16S rDNA gene, but other universal genes have been used in different fields of microbiology such as *rpoB*, *gyrB*, *pmoA*, and *cpn60* (*42, 43*). Different DNA regions or genes can be chosen as targets for molecular identification. For pathogenic organisms, virulence determinants are frequently used and often allow a very specific identification of bacterial species (*44*). Furthermore, by targeting virulence genes for PCR amplification, pathogenic strains can be discriminated from nonpathogenic strains from the same species (*45*).

In addition to gene detection, qualitative determination of virulence gene expression by techniques like reverse transcriptase-PCR (RT-PCR) can give important information on the viability or the pathogenicity of the cells (*46*). Some virulence determinants can give important information on the source (animal or human) of pollution. Different virulence determinants can be used to detect pathogenic strains (*47*). In the case of *E. coli* O157:H7, the *eaeA* gene is frequently used as it encodes intimin, an essential protein for attachment of the pathogen to epithelial cells as well as the genes encoding for the Shiga-like toxins I and II (*3, 44, 46-49*). It is important to note that many virulence determinants are located on plasmid DNA. To ensure reproducibility and reduce false negatives, when possible, chromosomally localized virulence genes are the preferred target due to plasmid instability and curing during lab manipulations.

1.3.1.2 mRNA Analyses

In general, mRNA (messenger RNA) degrades at a faster rate in killed bacteria than DNA does, with the overall persistence of mRNA in dead cells dependant on the method used to kill the bacteria (e.g., heat, ethanol, ultraviolet irradiation). Nevertheless, mRNA-based approaches are difficult to apply in environmental sample monitoring because of the complexity of the methods required to extract detectable levels of intact mRNA from small numbers of bacteria. In addition, because the detection of mRNA can be difficult because of its instability (*40*) and because it does not always provide phylogenetic information, different assays targeting rRNA (ribosomal RNA) have been developed.

1.3.1.3 rRNA and pre-rRNA Analyses

Because rRNA molecules are abundant in viable cells, they can be used as an indicator of the physiological state of the target cells, either directly (*50*) or via DNA microarrays (*51*). Nevertheless, the main limitation of this powerful approach is the stability of mature rRNA, which may remain detectable for hours or days after the death of a microbial cell (*52, 53*). In this regard, the use of precursor rRNAs has been shown to better reflect the nutritional status and potentially, the viability of *E. coli* cells (*52*).

1.3.1.4 Additional Methods Based on DNA Analysis Systems

Detection of specific organisms by PCR is often limited by several biases such as the abundance of the targeted sequence in the total DNA pool. In that way, various combinations of techniques have been used to increase the sensitivity and the specificity of nucleic acid-based methods. All are based on a specific characteristic of the target (i.e. the size of the target fragment, the relative position of restriction sites, the secondary structure of the double-stranded DNA fragments, and the sequence of A, T, G, C nucleotides in the fragment).

- **Denaturing-gradient gel electrophoresis (DGGE)** can separate fragments of similar lengths but which having different sequences. In fact, in fragments of similar sequence, DGGE has been shown to discriminate up to a single base pair substitution. Recently this technique was applied to the differentiation of *E. coli* populations from environmental freshwaters by means of sequence variations in a fragment of the β-D-glucuronidase gene (*54*). This method was also used to compare the bacterial community dynamics in liquid swine manure during storage via the analysis of 16S rDNA (*55*).
- **Random Amplified Polymorphic DNA (RAPD)** involves the identification of polymorphisms within the DNA of fecal bacteria. This method has been applied in bacterial source tracking by screening primers to find sets of polymorphisms that are either unique to fecal bacteria from a given source or occur in a given source to a large and predictable degree (*56, 57*).

- **Specific genetic markers** can be used to discriminate human and animal isolates. Bernhard and Field (*58*) identified two human-specific genetic markers and five bovine-specific genetic markers in fecal samples by amplifying 16S ribosomal DNA (rDNA) fragments from members of the genus *Bifidobacterium* and the *Bacteroides-Prevotella* groups.
- **Integrated cell culture/polymerase chain reaction (ICC/PCR)** developed by Reynolds et al.(*59*) allows the monitoring of infectious viruses in environmental samples. Direct examination of samples by PCR can result in false negatives, however, coupling viral PCR with a cell culturing step drastically reduced the number of false negative PCR amplifications.

Several variations of the standard PCR have appeared over the years and these have served to enhance detection methods of pathogenic microorganisms in water. Multiplex PCR allows several targets to be co-amplified in one PCR by combining or multiplexing primer pairs (*48, 60, 61*). A major disadvantage of PCR is that the amount of DNA sequence known for a given organism may be limited. The DNA sequence upon which the diagnostic test is based must be unique for the microbe. In water quality areas, PCR has often been used for the detection of *Enterobacteriaceae* in water (*62-66*). However, the use of PCR in the detection and identification of rare events, such as pathogenic bacteria or protozoa has been hampered by several factors that are directly related to difficulties in detecting small numbers of organisms in a complex matrix (*67*). Due to the large diversity of the microbial community in environmental samples, as well as in feces, PCR in combination with other methods, such as immunoseparation, is becoming widely used.

1.4 DNA Microarrays

1.4.1 General Features

DNA microarrays can be defined as an ordered array of polynucleotides immobilized on a solid support of limited physical dimensions such as those of a microscope glass slide (typically 25mm x 75 mm). The capacity of polynucleotides to form double helices allows the immobilized molecules, customarily referred to as probes, to hybridize with complementary nucleic acids, referred to as targets, from a sample to be analyzed. The hybridization conditions can be chosen to allow for various experimental parameters such as the nature and length of the probes, the degree of similarity expected between probes and targets and the desired stringency of the analysis. Some form of labeling, often via attachment of a fluorescent molecule, is performed on the target nucleic acids to allow the observer to identify those parts of the array where hybridization has occurred (*68*).

Microarrays come in two major types, oligonucleotide microarrays and amplicon microarrays. Each type has its advantages. Modern synthetic techniques for oligonucleotides allow their length to be varied at will between a few bases to over 100 bases. This is useful because short oligonucleotides allow for very specific hybridization, since even one mismatch will prevent pairing at stringent hybridization temperatures, while longer oligonucleotide designs permit the detection of gene families where significant sequence variation has occurred, in a manner similar to hybridization with PCR amplicons. Amplicon arrays, on the other hand, offer advantages in terms of simplicity, cost for complex arrays and the capacity to bind families of related genes. The longer length of amplicons relative to oligonucleotides makes the amplicon array less specific in terms of differentiating small differences in nucleic acid sequences; this may be an advantage when the purpose of an experiment is to detect a broad range of bacteria within a genus.

Useful reviews on the uses of microarrays in environmental microbiology (*69, 70*) and more specifically on their use for pathogen detection in environmental samples (*71*) are available for additional information.

1.4.2 Sample Preparation Prior to Microarray Analysis

Direct nucleic acid extraction techniques—in which cells are disrupted before separation from particulate material—have been developed and are generally more effective than techniques that separate the cells from the particulates before disruption. However, these extracts require additional purification before they can be effectively used in hybridization or polymerase chain reaction procedures. After purification, environmental DNA can be examined following reduction of fragment size and cloning to obtain functional data or sequence information, or the fragments can be labeled and hybridized to microarrays. Alternatively, the DNA and RNA can be separated and processed to detect specific genes or to evaluate function or organism relatedness at the microbial community level (*72*).

Messenger RNA (mRNA) is an excellent measure of cellular state and activity at any given time. Unfortunately, RNA is highly susceptible to enzymatic degradation. Moreover, the asynchronous growth of bacterial cells in many natural environments, which are often characterized by starvation conditions, coupled with a rapid turnover and relatively short life cycles of bacterial transcripts, makes them difficult to isolate in appreciable quantities (*73*).

1.4.2.1 Concentration of Cells

Direct isolation of nucleic acids from the environment is useful in several respects, including the estimation of total biomass, detection of specific organisms and genes, estimation of species diversity, and cloning applications (*74*). The isolation of nucleic acids requires that the microorganisms can be harvested from the environment. Presently, this separation can be performed by either by a combination of filtration and centrifugation or by an immunomagnetic separation (IMS) of specific microorganisms.

1.4.2.2 Filtration and Centrifugation

Generally, wastewater samples are concentrated by filtration or centrifugation of known volumes of sample, whereby the bacteria present are captured on a filter or in a centrifuged pellet. In the case of sewage, the utility of filtration using 0.45 μm pore size membrane is limited because clogging restricts the sample volume. An additional consideration for microarray work is that filtration on a membrane implies resuspension of the bacterial cells or nucleic acids from the membrane, a process with inherent variability. For that reason, centrifugation is often preferred. However, with centrifugation there will be some obvious limitations in the size of the sample that can be examined.

1.4.2.3 Nucleic Acid Extraction

The efficiency of the extraction determines the quantity, quality, and diversity of the extracted DNA and RNA. Extraction procedures vary depending upon the source of the sample and the subsequent use for which the DNA is intended. Even within the limited confines of nucleic acid extraction from activated sludge samples, it has been shown that the optimum composition and sequence of cell lysis treatments may vary greatly, depending on the nature of the sludge and should be specifically adapted for sludge originating from different sources (*75*).

- DNA can be extracted by direct lysis procedures, which disrupt the bacterial cells directly in the samples, or by indirect lysis procedures, which require an initial cell extraction step and may be less efficient than direct lysis treatment. A major drawback of the direct lysis method is that more PCR-inhibitory substances are extracted along with the DNA. In addition, the number and diversity of the direct lysis DNA extraction protocols are daunting and depend on the nature of the sample to analyze. Nevertheless these protocols usually include one or more of the following elements: physical disruption, chemical lysis, and enzymatic lysis. Four different physical disruption techniques have been described (*76*): freeze-thaw disruption, bead mill homogenization, ultrasonication and grinding under liquid nitrogen. Many chemical lysis procedures incorporate an initial enzymatic lysis step with enzymes such as lysozyme, proteinase K, achromopeptidase, or pronase E (*76*). The chemical lysis then proceeds with buffers that may contain various detergents (sodium dodecyl sulfate, Sarkosyl, Triton X-100) and/or chaotropic agents such as guanidinium thiocyanate. Common modifications of the basic chemical lysis include high-temperature incubation, a phenol or chloroform extraction step and incorporation of chelating agents (EDTA, Chelex 100) to inhibit nucleases and disperse particles. Removal of inhibitory substances that interfere with PCR is performed through additonal steps such as purification of the crude DNA on Sephadex G-200 and silica-based SpinBind column purifications (*76*).
- There are also many commercial DNA purification kits with proprietary columns and buffers of undisclosed composition. Suitability of these kits must be ascertained for each type of sample under consideration.

1.4.2.4 DNA Fragmentation

When total DNA from a culture or environmental sample is to be applied to a microarray, it is generally considered essential to fragment the high molecular weight DNA to a smaller size. This may be achieved by restriction enzymatic digestion or by sonication. A method combining DNA fragmentation based on free radical chemistry with fluorescent dye labeling has been reported (*77*). The optimal level of DNA fragmentation remains unclear. Wilson et al. (*78*) reported that hybridization is enhanced with fragments size *ca.* 50 bp. This result may be specific to either to the high-density oligonucleotide array used or to the labeling method used (terminal transferase), inasmuch as other researchers (*79*) obtained better signals with fragments in size ranges above 1 kb.

1.4.2.5 Labeling and Detection Methods

Microbiology has generally relied on the isolation of organisms in pure culture, followed by identification and characterization of each isolate. In contrast, the microarray approach has the potential to analyze, in a single experiment, the total extracted RNA or DNA on an array that detects all the species that may reasonably be expected in the sample. This parallel processing power could allow experimental designs that would be much less costly and time-consuming than conventional methods. The most common approach to detection and measurement of hybridization between the sample and the relevant probes on the microarray is to label the DNA with a covalently attached fluorescent molecule. Direct fluorescent enzymatic labeling of DNA extracted from environmental samples presents difficulties because of possible co-purification of polymerase inhibitors such as humic acid, porphyrins, and colloidal iron. Moreover, nucleotide triphosphates carrying covalently attached fluorescent dyes have a poor efficiency of incorporation by DNA polymerases. Alternate labeling approaches, such as direct chemical labeling, are therefore favored in many applications, and there are several commercially available reagents for this purpose. Some researchers also use indirect labeling, where an amino allyl modified nucleotide is incorporated into DNA by a DNA polymerase and subsequently reacted with an activated flurorescent dye. There is no consensus at this time as to which method may be generally superior and further development will undoubtedly occur (*80*).

1.4.3 Current Status of Microarrays in Microbial Detection

1.4.3.1 Sensitivity and Specificity

A current limitation of microarrays is that they do not offer the low detection levels available by PCR and may not be suitable for detection of low abundance organisms, such as *Cryptosporidium* in biologically complex samples. Reported detection limits are in the range of 10^7 molecules of a given target in the applied sample (*51, 61, 81*), depending somewhat on the type of microarrays, samples and labeling methods used. Microarrays are currently estimated to be between 100- and 10,000-fold less sensitive than PCR and between 10 and 1,000-fold less sensitive than conventional membrane-based hybridization (*72, 82*).

In order to achieve high sensitivity, the use of 16S rDNA as a target gene for bacterial detection has obvious appeal. The rRNA molecule is abundant (from 10^3 to 10^5 per cell, depending on growth status) and has been broadly used as a taxonomic tool, with extensive databases available. However, there are also problems to be anticipated, such as the extensive secondary structure of rRNA, which may create hybridization problems. A published study by Small et al. (*51*) addresses some of these issues; in particular, the use of 'chaperone' oligonucleotides to open up the 16S structure for hybridization was found to be essential. Detection limit was estimated at 0.5 μg of total RNA, representing approximately 7.5 x 10^6 cells.

With regards to detection of functional genes, the work of Wu et al. (*82*) is of interest. Their work showed that the volume of sample applied to the microarray has a profound effect on detection limit, with a 2 μl volume giving a 500-fold increase in sensitivity over the more conventional 15 μl volume. They have used an amplicon array with a detection limit of approximately 1 ng of genomic DNA from the organism containing the nitrite reductase gene *nirS*. This translates to detection of about 3 x 10^5 organisms in pure culture. Also noteworthy is the recent work of Dennis et al. (*83*) on the detection of mRNAs from bacterial communities via hybridization of their cDNAs to functional gene microarrays. Their results show good differentiation between the target organism and the community background and open the possibility of assessing viability through assessment of the ability of a target species to induce new mRNA synthesis in presence of a stimulant such as heat shock.

In situations where detection of a given pathogen is more important than its quantitation, the coupling of PCR (for low detection limit) with microarrays (for confirmation of the identity of the amplicon) can be quite useful (*71*). It is also possible to go further and to use a prior step of immunomagnetic concentration of the target organisms (*61*), followed by PCR and microarray to confirm presence of low amounts of a pathogen.

The main problem with using PCR prior to microarray analysis is that PCR is known to suffer from bias (*84, 85*). In particular, the transition from the exponential phase to the plateau phase occurs earlier for the more abundant species present in the sample. There are also biases related to G+C content of the amplicons and to variable priming efficacy from 'universal' primers. As a result, the quantitation potential of microarrays is greatly diminished or abolished if a prior step of PCR is involved in sample preparation. This situation is illustrated by the work of Wilson et al.(*78*), using a high-density, high-redundancy Affymetrix microarray carrying 31179 oligonucleotide probes for over 1,900 prokaryotic 16S sequences. The authors noted that there was not a good agreement between the frequency of a given 16S sequence as determined by cloning and sequencing the PCR amplicons and the intensity of the corresponding spots on the microarray.

The use of PCR in sample preparation, prior to microarray hybridization, often requires additional processing steps to remove known PCR inhibitors. It is also important to note that PCR biases may be aggravated or mitigated by differences in cell, spore or oocyst lysis and by variations in DNA extraction efficiency; at this point in time, quantitative statements made on the basis of microarrays used for detection of microorganisms must be viewed with caution (*72*).

1.4.3.2 Taxonomic Versus Pathotype Microarrays

When dealing with bacterial species such as *E. coli* that encompass both pathogenic and non-pathogenic strains and for which many different pathotypes are found, it can be helpful to use microarrays to determine these pathotypes (*79, 86*). Microarrays can also combine taxonomical and antibiotic genotype identification such as found in the work of Gingeras et al. (*87*). In this publication, the authors used Affymetrix microarrays to characterize mycobacterial isolates both in terms of their taxonomy and their resistance to rifampin. The ability to sequence 16S and *rpoB* by hybridization provided highly detailed information on the identity, population structure, and antibiotic resistances of the strains under study.

1.4.4 Applications Relevant to Wastewater Analysis

1.4.4.1 Use of Microarrays to Characterize Microbial Communities in Coastal Water

Microarrays show promise for the characterization of complex microbial communities, including organisms requiring special culture conditions as well as non-culturable types. These characteristics are particularly applicable to microbial communities in marine samples, as these are known to contain a high proportion of non-culturable organisms. Recently, Stine et al. (*88*) described the development of such a microarray designed to characterize microbial communities in coastal waters in the Chesapeake Bay area.

1.4.4.2 Use of Microarrays for Detection of Fecal Pollution Indicators

A recent publication by Cerniglia's group (*89*) at the United States Food and Drug Administration used conserved primers to PCR amplify 16S genes, followed by microarray analysis to detect the presence of 20 intestinal bacteria predominantly found in humans. While this approach suffers from the bias limitations of PCR discussed earlier, it nevertheless points the way towards subsequent development of microarrays capable of tracing fecal pollution back to human or animal sources. The work was performed on fecal samples and further development will be needed to develop the technique for wastewater samples. This research offers a possible approach for the differentiation between human and animal fecal contamination.

1.4.4.3 Use of Microarrays to Monitor Genetic Variability in Pathogens

There is one application in water analysis where microarray technology is currently functional, the area of genotyping microorganisms for the purpose of epidemiology or source tracking. The work published by Straub et al. (*90*) provides an excellent illustration where single nucleotide polymorphisms in the conserved *hsp70* gene are used to identify *Cryptosporidium* species. In this situation, where quantification is not a consideration, PCR amplification is eminently suited to generate the target DNA for subsequent microarray hybridizations, while the use of immobilized oligonucleotides allows for single base differentiation of polymorphisms.

Note: An expanded version of this chapter has been accepted for publication (Lemarchand et al.) in the August 2004 issue of *Critical Reviews in Microbiology*. Readers are referred to this expanded version for additional information.

CHAPTER 2.0

OBJECTIVES OF THE ORIGINAL PROPOSAL

2.1 Background

Natural waters are subject to sewage or wastewater discharges as a result of human activity or stormwater runoff. These effluents contain a wide range of microorganisms (viruses, bacteria, and protozoa) that can be waterborne pathogens. Most waterborne pathogens are difficult to detect and enumerate with accuracy primarily due to methodological limitations (e.g. their low concentrations compared to non pathogenic microorganisms). In addition, the accurate detection of waterborne microorganisms implies high money and time costs. This project is being done to help bring DNA microarray technology into use for the detection of important waterborne pathogens in wastewater treatment plant effluents. When compared to polymerase chain reaction (PCR), the molecular method most widely used in pathogen detection, DNA microarrays offer potential advantages in the ability to detect hundreds of possible pathogens simultaneously. This advantage is currently offset by the low sensitivity of current methodology and by the extensive sample processing needed upstream of the microarray hybridization step.

As reported in the bibliographic review, the quantitative detection limits of microarrays, in terms of number of cells detected in environmental samples, are unknown. Nevertheless, it is crucial to determine these limits before applying DNA microarrays in water quality control assessments. In addition, different physicochemical parameters can interfere with DNA extraction in wastewaters and must be considered.

2.2 Objectives

The objectives of the original proposal were to:

- Perform a literature review to guide the project;
- Develop an index of fecal pollution using DNA microarrays;
- Develop DNA microarray methods to distinguish between fecal pollution of human origin versus animal origin;
- Explore the use of DNA microarrays as indicators for efficacy of treatment; and
- Develop DNA microarray methods for detection of pathogenic microorganisms in water.

The literature review indicated that the use of microarray as indicators for efficacy of treatment was unpromising, inasmuch as nucleic acids can persist for hours or even days after death of bacterial cells, depending on the biocidal agent used. Additionally, the literature review highlighted the need for a comprehensive study of DNA extraction protocols. The work plan was therefore built in four parts, each of which constitutes a separate chapter in this report:

Chapter 3.0:	Sample Collection, Basic Microbiological Analysis, and Development of a DNA Extraction Method Suitable for Microarray Analysis
Chapter 4.0:	Development and Evaluation of Microarrays for Pathogen Detection
Chapter 5.0:	Development and Evaluation of Microarrays for Bacterial Source Tracking
Chapter 6.0:	Considerations on the Usability of DNA Microarrays in Wastewater Analysis

CHAPTER 3.0

SAMPLE COLLECTION AND DEVELOPMENT OF A DNA EXTRACTION METHOD SUITABLE FOR MICROARRAY ANALYSIS

3.1 Description of the Sample Collection Procedures

3.1.1 Human Wastewater Samples

Raw wastewater samples were collected from the Fabreville wastewater treatment plant (Laval, Quebec). This plant was built in 1984 and utilizes a physico-chemical process. Twenty liters of wastewater was concentrated as described in the standard wastewater sample concentration procedure (Appendix A).

3.1.2 Animal Contaminated Water Samples

The water samples contaminated with animal manure were supplied by IRDA (Institut de Recherche et de Développement en Agroenvironnement, 3300, rue Sicotte, St-Hyacinthe, Québec, J2S 7B6).

The wastewater samples were directly collected from underground drains located under a manure-treated soil. Bovine and porcine manures were spread separately on different soils at the end of October 2003, and drainage water was collected after the first rainy periods. Two 250-mL samples of bovine drainage water and one 200 mL sample of porcine drainage water were collected. The bovine and porcine drainage water samples were concentrated as described in the Standard wastewater sample concentration procedure (Appendix A). In contrast to the 900 mL aliquots concentrated for the human wastewater samples, bovine contaminated water samples were divided into five aliquots of 115 mL and concentrated to obtain a volume of 8 mL. The porcine contaminated water was divided into 13 mL aliquots without the previous concentration steps.

3.2 Bacterial Cell Counts

Bacterial cell concentrations in raw water were determined by culture on Plate Count Agar (PCA) (http://www.bioconnections.co.uk/cataloguepages/info_pages/e-p/BC2010.htm) for heterotrophic counts as described in the Standard heterotrophic bacteria count on PCA agar procedure (Appendix B). *E. coli* counts were determined by culture on mTEC agar as described in the Standard *E. coli* count on mTEC agar procedure (Appendix C).

The mTEC method is recommended for use in monitoring fresh, estuarine, and marine surface waters. It was specifically developed to be used as a measure of recreational water quality. The mTEC agar contains selective and differential agents. Sodium lauryl sulfate and sodium deoxycholate inhibit Gram-positive cocci and endospore-forming bacteria. Bromocresol purple and bromophenol red inhibit non-target bacteria and cause a color change from purple to yellow when lactose is utilized. Lactose utilization (yellow color) is the basis for presumptive identification of *E. coli*. For the second step, after incubation on mTEC, filter membranes containing yellow colonies were transferred to pads saturated with urea-phenol red solution. Colonies that are not *E. coli* turn from yellow to purple. This color change is caused by the breakdown of urea by the enzyme urease and a subsequent increase in pH. *E. coli* is urease negative, and colonies of *E. coli* remain yellow.

PCA medium does not contain any inhibitors or indicators; it is mainly used to determine the total microbial content in milk, dairy products, water, and other materials. The PCA agar procedure is used to enumerate the heterotrophic bacteria.

3.3 DNA Extraction Procedure Comparison

Direct isolation of nucleic acids from the aquatic environment may be useful in several respects, including the estimation of total biomass, detection of specific organisms and genes, estimations of species diversity and cloning applications. The isolation of DNA is one of the more commonly used procedures in genetics, molecular biology, and microbiology. Purified DNA is required for many applications such as gene cloning, PCR, and DNA hybridizations. The availability of effective DNA extraction methods is essential for environmental microbiology studies. Several protocols have been developed to obtain contaminant-free DNA directly from an environmental matrix. These techniques are mostly based on enzymatic or mechanical cell lysis.

The general principles are:

1. Lysis of the cells
2. Solubilization and extraction of the DNA
3. Precipitation of the DNA; and
4. Use of chelating agents, detergents, and chemical denaturants to avoid enzymatic digestion of the DNA by endogenous nucleases.

After a literature review and a consideration of different physicochemical parameters, 10 procedures were compared:

1. Guanidine thiocyanate [GuSCN]-diatomaceous earth [DE] extraction *(91)*
2. Disruption of cells by sonication followed by a phenol/chloroform/isoamyl alcohol (PCI) extraction
3. Lysis of cells by guanidine buffer followed by a PCI extraction [as suggested in Qiagen genomic DNA buffer set handbook (Cat No. 19060)]
4. UltraClean Water DNA isolation kit (MO BIO laboratories, Inc) following manufacturer instructions: MoBio Laboratories, Inc UltraClean Water DNA isolation kit (Cat No.14800-10)
5. DNA extraction by using GES procedure [500 mM guanidium thiocyanate, 100 mM EDTA, 0.5% (w/v) N-lauryl-sarcosine] *(92)*
6. Triton-Prep method (from U.S. Department of Commerce http://micro.nwfsc.noaa.gov/protocols/tritonprep.html)
7. Freeze-thaw and lysozyme-sodium dodecyl sulfate (SDS) lysis procedures (adaptation of the technique described by Bourrain et al. *(93)*
8. Bead-beating disruption and ammonium acetate purification *(94)*
9. Bead-beating disruption and phenol/chloroform/isopropanol extraction *(95)*

10. Bead-beating disruption, potassium acetate impurity precipitation, and phenol/chloroform/isoamylalcohol extraction (*96*)

A summary of the 10 different procedures is shown in Table 3-1.

3.3.1 Choice of DNA Extraction Technique for Microarray Hybridization Tests

Four parameters (DNA quantity, DNA purity, DNA degradation, and absence of PCR inhibitors) were analyzed to determine the efficiency of each DNA extraction technique in addition to the quality of the extracted DNA. A comparison of these parameters was executed in order to determine the appropriate technique to use in our hybridization study. This comparison is summarized in Table 3-2.

After this analysis, it was concluded that only procedures 4, 7, and 8 completely satisfied our selection criteria. However, because results were similar for the three techniques and the cost is lower for procedures 7 and 8 in comparison to procedure 4 (MOBIO DNA extraction kit), where many reactions of the commercial kit must be used for a single wastewater pellet, we decided to focus on procedures 7 and 8. We focused on procedure 8 for the following reasons. First, procedure 8 required fewer manipulation steps than procedure 7 and, therefore, the probability of losing some DNA was reduced. Second, the length of procedure 8 (about 4 hours) was about half that of procedure 7 (about 7 hours). Third, because all the manipulations of procedure 8 need small volumes and can be done in standard Eppendorf microtubes, a common laboratory microcentrifuge can be employed. Therefore, this procedure could be performed in the majority of molecular microbiology laboratories.

3.4 Conclusions about the DNA Extraction Procedure for Microarray Hybridization Tests

In conclusion, we decided to use procedure 8 (bead-beating separation and ammonium acetate purification) to extract genomic DNA from the water samples contaminated with human or animal microorganisms. The standard wastewater DNA extraction procedure is described in Appendix D.

Table 3-1. Summary of the Different Steps of the 10 Extraction Procedures Tested.

	Types of Cell Lysis				General Purification Steps
Procedure #	*Physical*	*Description*	*Chemical*	*Description*	*Description*
1	✓	**Buffer** (800 mM Guanidine, 30 mM TrisHCl, 5% (v/v) Tween-20, 0.5% (v/v) Triton X-100) **and Proteinase K** (1 mg/mL)			-Diatomaceous earth purification
2	✓	**SFM Buffer** (10 mg lysozyme, 0.5M NaCl, 0.01M EDTA), **4% (w/v) SDS and 1 mg/mL Proteinase K**	✓	Sonication	-PCI purification -Isopropanol and ethanol precipitations
3	✓	**Buffer** (800 mM Guanidine, 30 mM TrisHCl, 5% (v/v) Tween-20, 0.5% (v/v) Triton X-100) **and Proteinase K** (1 mg/mL)			-PCI + ammonium acetate purifications -Isopropanol and ethanol precipitations
4	✓	Unknown composition of the buffers (commercial kit)	✓	Bead beating	-Commercial columns
5	✓	**GES Buffer** (500 mM Guanidium thiocyanate, 100 mM EDTA, 0.5% (w/v) N-lauryl-N-Sarcosine)			-PCI purification -Isopropanol and ethanol precipitations
6	✓	**STET Buffer** (8% (w/v) Sucrose, 5% (v/v) Triton X-100, 50 mM Tris-HCl, 50 mM EDTA, 1 mg/mL lysozyme)			-LiCl purification -Isopropanol and ethanol precipitations
7	✓	**TENP Buffer** (50 mM Tris base, 20 mM EDTA, 100 mM NaCl, 0.01 g/mL polyvinylpyrrolidone), **lysozyme, 5% SDS**	✓	Freeze-thawing	-PCI purification -Isopropanol and ethanol precipitations
8	✓	**Buffer** (50 mM Tris-HCl, 5 mM EDTA, 3% (w/v) SDS)	✓	Bead beating	-Ammonium acetate purification -Isopropanol and ethanol precipitations -PVPP column purification
9	✓	**Buffer** (50 mM Tris-HCl, 50 mM EDTA, 3% (w/v) SDS)	✓	Bead beating	-PCI purification -Ethanol precipitation
10	✓	**Buffer** (100 mM Tris-HCl, 100 mM EDTA, 1.5 M NaCl) **and 2% (w/v) SDS**	✓	Bead beating	-PCI + Ammonium acetate -Isopropanol and ethanol precipitations

Table 3-2. Comparison of All DNA Extraction Procedures Used in Our Study.

Procedure	DNA Quantity	DNA Purity	DNA Degradation	PCR Amplification
1 Chemical lysis and DNA capture by a diatomaceous earth matrix	+++	-	+++	+
2 Sonication + PCI extraction	+++++	+++++	+++	+
3 Chemical lysis + PCI extraction	++	+++	+	+
4 UltraClean Water DNA isolation Kit (MO BIO)	+++++	++	+	+
5 GES procedure	+++++	+ and PR	+	+
6 Triton-prep method	-	-	N.D.	-
7 Freeze-thaw (55°C) and lysozyme-SDS lysis procedure	++++	+++	++	+
8 Bead-beating disruption and ammonium acetate purification (*94*)	+++++	++++	++	+
9 Bead-beating disruption and PCI extraction	+++++	++ (Brownish)	++	+
10 Bead-beating disruption, potassium acetate impurity precipitation and PCI extraction	++	+	++	+

Note 1: The gray cells represent unsatisfactory results.
Note 2: The results presented in this table represent the means of duplicate experiments.

Legend:

DNA quantity:
- ; 0 to 5 μg
+ ; 5 to 19 μg
++ ; 20 to 39 μg
+++ ; 40 to 59 μg
++++ ; 60 to 79 μg
+++++ ; 80 μg and more

DNA purity:
- ; A_{260}/A_{280} below 1.5 (impure sample)
+ ; A_{260}/A_{280} = 1.5 to 1.6
++ ; A_{260}/A_{280} = 1.6 to 1.7
+++ ; A_{260}/A_{280} = 1.7 to 1.8
++++ ; A_{260}/A_{280} = 1.8 to 1.9
+++++ ; A_{260}/A_{280} = 1.9 to 2.0
PR= Phenol residues

DNA degradation:
- ; No degradation
+ ; Low degradation (fragment sizes between 2 kb and 23 kb)
++ ; Medium degradation (fragment sizes between 500 bp and 23 kb)
+++ ; High degradation (fragment sizes less than 500 bp)
N.D.; Not determined

CHAPTER 4.0

DEVELOPMENT AND EVALUATION OF MICROARRAYS FOR PATHOGEN DETECTION

4.1 Total *Escherichia coli* DNA on 50-mer Oligonucleotide Versus Amplicon Arrays

The microarray prototype combines 50-mer oligonucleotides and longer, denatured double-stranded DNA (amplicons) on the same array to compare the signal intensity obtained by these two different types of probes.

4.1.1 Design

Initially, it was deemed desirable to design oligonucleotide probes for different genes of *E. coli* to allow the specific detection of this microorganism in complex environments. As it is difficult to find a single gene for this purpose, a combination of four genes was used on our prototype microarray:

- *uidA:* (synonyms: *gurA*, *gusA*), coding for ß-glucuronidase
- *fimA:* coding for a major type 1 subunit fimbrin (pilin), present in all *E. coli* strains
- *cpn60:* (alternate gene names *groE*, *groEL*, *hdh*, *tabB*, *mopA*), coding for *E. coli* chaperone Hsp60
- 16S rDNA: coding for the *E. coli* 16S ribosomal subunit

The simultaneous detection of these genes allowed us to confirm the specific presence of *E. coli* cells in the samples. To control the validity of the method a positive control was added to each array corresponding to the reporter GFP gene inserted in the spiking control (Table 4-1).

Five amplicons were added to the initial pattern to allow a comparison between the signals obtained by these two kinds of probes. The target genes were the same as those targeted by 50-mer oligonucleotides and were obtained by PCR amplification. The sequences of the primer sets used in this study are given in Table 4-2. The annealing temperatures were variable, according to the melting temperature (Tm) of each primer set. All amplifications were based on the following scheme:

Hot start: 3 min at 94°C
30 cycles of amplification: 1 min at 94°C
1 min at calculated annealing temperature
1 min at 72°C
Final extension: 7 min at 72°C

The pattern of the microarray prototype 1.0 is presented in Figure 4-1.

Table 4-1. Sequences and GC Content Percentages of the 50-mer Oligonucleotides Designed for Prototype 1.0.

Gene	Oligonucleotide Sequence	G+C %
uidA	TAGAGCATTACGCTGCGATGGATCCCGGCATAGTTAAAGAAATCATGGAA	44
fimA	GAATGTTAAAACCGACAGCAGAACTGGTTGCTCCTTCCTGTGCCAGCGAT	50
Cpn60	CACGGTACGGACAGCGCTTTCAGTTCTTCAACTGCAGCGGTAACTGCTTT	52
16S rDNA	TGCGGGTAACGTCAATGAGCAAAGGTATTAACTTTACTCCCTTCCTCCCC	48
*Gfpmut*3.1 (*93*)	GGGCAGATTGTGTGGACAGGTAATGGTTGTCTGGTAAAAGGACAGGGCCA	50
*Gfpmut*3.1 (*95*)	GGTCTCTCTTTTCGTTGGGATCTTTCGAAAGGGCAGATTGTGTGGACAGG	50

Table 4-2. Sequences of Primer Pairs and Size of the Resultant Amplicons.

Gene	Forward Primer	Reverse Primer	Size (bp)	Reference
uidA	GCGAAAACTGTGGAATTGGG	TGATGCTCCATAACTTCCTG	250	(*97*)
fimA	GTTGATCAAACCGTTCAG	AATAACGCGCCTGGAACG	331	This project
cpn60	CTGCAGGCGACGGTACCACCACTG	TCGCCGAAGCCCGGTGCTTTAACC	604	This project
16S rDNA	GAGTTTGATCCTGGCTCAG	G(TA)ATTACCGCGGC(TG)GCTG	528	(*98*)
*gfpmut*3.1	ATGAGTAAAGGAGAAGAACTTTT	GCTATTTGTATAGTTCATCCATGCC ATGTGT	717	This project

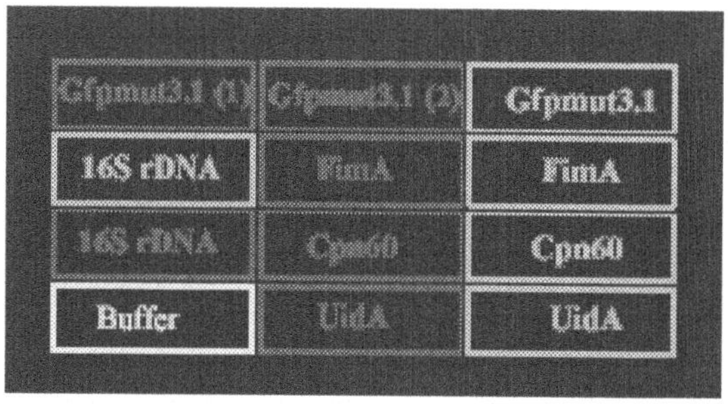

Figure 4-1. Probe Pattern of the Microarray Prototype 1.0. Amplicons Are Presented in Blue Characters, 50-mer Oligonucleotides in Red, and the Negative Control in Yellow Characters.

4.1.2 DNA Labeling Procedure

Fragmented *E. coli* genomic DNA was employed for the hybridization. The fragmentation was done using a Virsonic 100 ultrasonic cell disrupter (VirTis) at power 3 during three seconds to disrupt 10µg of purified genomic DNA. Genomic DNA was labeled using the Mirus' Cy3 Label IT Nucleic Acid Labeling kit according to the manufacturer's instructions and incubated three hours at 37°C in the dark. As the labeling reaction was performed in a small volume (30 µl), a quick spin (i.e., 2 seconds at 12,000 rpm) was performed after 30 minutes' incubation to minimize the effects of evaporation and keep the concentration of the reaction components at the appropriate levels. The unreacted labeling reagents were removed from the labeled nucleic acid using either a QIAquick PCR purification kit (Qiagen) or a Microcon YM-30 (Millipore).

4.1.3 Microarray Hybridization

Microarrays were prehybridized at 42°C for one hour under a 9x9 mm coverslip (Sigma) in 2 µL of pre-warmed solution A (DIG Easy Hyb buffer, Roche, containing 10 µg of denatured salmon sperm DNA). After the coverslip was removed by dipping the slide in 0.1X SSC (1X SSC contained 150 mM NaCl and 15 mM trisodium citrate, pH 7), the array was rinsed in water and dried by centrifugation at room temperature in 50 mL conical tubes for 5 min at 150 g. Fluorescently-labeled DNA was chemically denatured as described by the manufacturer and added to 2 µl of fresh pre-warmed solution A. Hybridization was carried out overnight at 42°C as recommended by the manufacturer. After hybridization, the coverslip was then removed in 0.1X SSC and the microarray washed three times in prewarmed 0.1X SSC/0.1% SDS solution and once in 0.1X SSC for 10 min at 50°C. After drying by centrifugation (150 g, five min, room temperature), the array was analyzed using a fluorescent scanner (ScanArray Lite, Canberra-Packard, Mississauga, Ontario).

4.1.4 Results: Prototype 1.0 Detection Limit

The detection limits of the prototype 1.0 correspond to 100 ng of specific genomic DNA (Figure 4-2). With 50 and 25 ng of labeled DNA, no signal was observed.

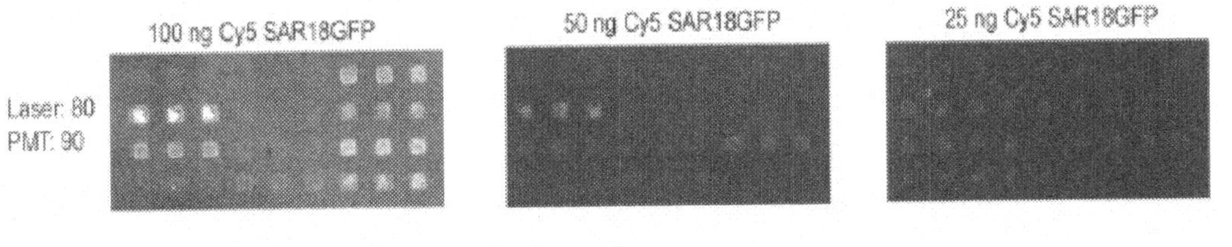

Figure 4-2: Hybridization of 100 ng, 50 ng, and 25 ng of *E. coli* SAR18-GFP-Cy5 DNA.

4.1.5 Conclusions

- The lowest detectable limit of labeled genomic DNA is 100 ng of *E. coli* genomic DNA corresponding to 2 x 10^7 genomes. This estimate is based on the genome size of *Escherichia coli* K12 strain MG1655, as follows:

Genome size = 4.64 million bases, GenBank file NC_000913

One base pair = 650 Daltons molecular weight (average)
Molecular weight of one genome = 650 Daltons x 4.64 x 10^6 = 3.016 x 10^9 grams per mole
Weight of one genome = molecular weight/Avogadro's number = 3.016 x 10^9/ 6.023 x 10^{23} = 5 x 10^{-15} grams = 5 fg
Therefore, 50 ng = 10^7 genomes.

- The amplicons are more sensitive than oligonucleotides.
- The direct use of genomic DNA as target on microarrays is a great challenge but it currently requires the presence of a minimum of 2 x 10^7 targeted genomes in the sample.
- Due to their normally low numbers in water, the detection of pathogens in water requires increased sensitivity. To overcome this problem, we decided to design a new microarray prototype based on PCR amplification of the target. This will be shown in the subsequent sections.

4.2 16S rDNA and *cpn60 Escherichia coli* Amplicons on 70-mer Oligonucleotide Versus Amplicon Arrays

As previously shown, hybridization of amplicons on microarrays gives high intensity signals (*99*). However the use of amplicon targets is not suitable when the targets are specific genes that require a specific set of primers and, therefore, specific PCR amplification conditions. The only way to avoid the need of specific amplification conditions is to target conserved sequences (such as 16S rDNA and *cpn60* genes) with universal primers. To test the hybridization of amplicons on microarrays, we designed a prototype 2.0 combining 70-mer oligonucleotide and amplicon probes. These probes were specific for the 16S rDNA and *cpn60* genes of different waterborne pathogens.

4.2.1 Design

In this prototype, 70-mer oligonucleotides were designed to increase the strength of hybridization in terms of increased signal strength. To compare hybridizations on an amplicon microarray versus an oligonucleotide microarray it is important to be sure that:

- No competition occurs between oligonucleotide and amplicon probes; and
- The absence of hybridization signals is directly related to the absence of the targeted sequence (i.e. no false negatives).

To address the first point, the arrays for oligonucleotides and for amplicons were printed as subarrays on physically distant regions of the microarray slide and were hybridized under separate cover slips. To address the second point, two or three different 70-mer oligonucleotide probes were designed for each target species when possible.

The OligoPicker software is designed to select up to five oligonucleotide probes for each of the DNA sequences provided for microarray spotting (*100*). OligoPicker picks specific oligonucleotides by skipping regions with contiguous bases common in other sequences. The sequences of the selected oligonucleotides are presented in
Table 4-3.

The amplicon probes were made by PCR amplification. The sequences of the primer sets used for these amplifications are given in Table 4-4. The annealing temperatures were variable according to the melting temperature (Tm) of each primer set. All amplifications were based on the following scheme:

- ***Hot start***: 3 min at 94°C
- ***30 cycles of amplification***: 1 min at 94°C
1 min at calculated annealing temperature
1 min at 72°C
- ***Final extension***: 7 min at 72°C

The pattern of the microarray prototype 2.0 is presented in Figure 4-3.

Table 4-3. Sequences of 70-mer Oligonucleotides Selected for Each Target Microorganism.

gene	oligonucleotide #1 (5'-3')	oligonucleotide #2 (5'-3')	oligonucleotide #3 (5'-3')
Campylobacter jejuni cpn60	tgcccacttttccatagcatcagcgattaaatttccga tttttcatcagagttggctgagattgtagc	cccctttgaaatattaagaacaccgcgaagtttattt acaaccaaagttgcaagcgcttcaccttcaata	
Campylobacter jejuni 16S	gttaagcaggagtatagagtattagcagtcgtttccaac tgttgtcctcttgtgtagggcagattaacta	ttagccggtgcttattccttaggtaccgtcagaattc ttccctaagaaaaggagtttacgctccgaaaag	
Legionella pneumophila cpn60	catcttctgcgatgatcaataaaggacggccagattttg caacaccttccaatacggacaacatttcacg	cttcatcggaattagcagaaatagttccaacttgagc aatagctttgctgtctttgcatggcttagacat	
Legionella pneumophila 16S	tgcttcttctgtgggtaacgtccagttaatcagctctta acctatcaaccctcctccccactgaaagtgc	tgtccccctcttcaaggcatattcctacgcgttactc acccgttcgccactcgccatctgtctagcaagc	
Pseudomonas aeruginosa cpn60	tggccagctctttcagctgggcgacgatggccacggtgg ccttgtcgatgccacgcttcaggtccatcgg	cagcagcagcgggctatccagctcggcagccatggtg tccggcttgttcacgaagtagggggacaggtag	tgccttcgacgacggacagttcgttttccaggcccgag ccttcctcgacggtgatcacgccttctttacc
Pseudomonas aeruginosa 16S	ataccgcatacgtcctgagggagaaagtggggatcttc ggacctcacgctatcagatgagcctaggtcg		
Vibrio cholerae cpn60	gcttcgccttccacatcttcagccacgatcagcagtgga cgagaggctttcgctacgccttccagtactg	gccgtcacgacccactttttccatcgcttcagcaatg atgttgcccacgctagagtctgagttagcagag	tcagcaatgatgttgcccacgctagagtctgagttagc agagatggtacctacttgggcaatcgctttag
Vibrio cholerae 16S	tgcggtattagccatcgtttccaatggttatccccctct accgggcaatttcccaggcattactcacccg		
Yersinia enterocolitica cpn60	tccatcgcttgtgcaatcagctcacccacagttgagtct gagtttgcagagatggtccctacctgagcaa		
Yersinia enterocolitica 16S	cttcttctgcgagtaacgtcaatcacaaaggttattaac ctttatgccttcctcctcgctgaaagtgctt	tctgggcacatccgatggcgtgaggcccgaaggtccc ccactattggtccgaagacgttatgcggtatta	
Aeromonas hydrophila cpn60	ttgttgtcagcacacggctgggacagcgcttgcagctcg gcaacggcggccactacggccttgtcgatac	agatggtgccgacctgagcgatggcattgttgtcagc acacggctgggacagcgcttgcagctcggcaac	
Aeromonas hydrophila 16S	tatctgctggctgtgacgttactcgcagaagaagcaccg gctaactccgtgccagcagccgcggtaatac		
Escherichia coli O157:H7 *cpn60*	gctttagagtcagagcacggtacggacagcgctttcagt tcttcaactgcagcggtaacagctttgtcga	agcgcttcgccttctacatcttcagcgatgatcagca gcggtttgcctgctttggcaacggcttccagaa	
Escherichia coli O157:H7 16S	gtgcttcttctgcgggtaacgtcaatgagcaaaggtatt aactttactcccttcctccccgctgaaagta	cttcttctgcgggtaacgtcaatgagcaaaggtatta actttactcccttcctccccgctgaaagtactt	
Arcobacter sp. *Cpn60*	ccagtgttgctaatgcttcaccatctacatcttcagcaa taattactaaaggtcttcctgattgatttac	ctatttactaccagtgttgctaatgcttcaccatcta catcttcagcaataattactaaaggtcttcctg	
Arcobacter sp. 16S	cgtcatcgtcttggtaggccattaccccaccaactaact gatacaatacaggctaatctcttaccaataa	catataataccgtcattatcttcttatataaaggag tttacgcaccgaaatgtgtcatcctccacgcgg	tactcacccgtgcgccacttagctgacaattatagcaa gctataatccgttctcgttcgacttgcatgtg

Table 4-4. Primers Sets Used to Generate 16S rDNA and *cpn60* Amplicons for Microarray Printing Prototype 2.0.

Gene	Forward Primer	Reverse Primer	Annealing T°C	Reference
16S rDNA (universal primers)	GAGTTTGATCCTGGCT CAG	G(T/A)ATTACCGCGGC(T/G) GCTG	58	(*98*)
Legionella pneumophila cpn 60	GCAACAGTATTGGCTC GTTCTATT	GACAGCACATACTTTTACA ATACC	48	This project
Escherichia coli O157:H7 *cpn60*	GCAACCGTACTGGCTC AGGCTAT	AACTGCAGCAACTTTCACG ATGCC	50	This project
Salmonella typhimurium cpn60	GCGACCGTACTGGCG CAGTCCAT	AACCGCAGCCACTTTCAC GATGCC	56	This project
Aeromonas hydrophila cpn60	GCTACCGTACTGGCTC AAGCCATC	GACGGCAGCGACTTTCAC GATGCC	60	This project
Pseudomonas aeruginosa cpn60	GCGACCGTCCTGGCCC AGGCCAT	GACAGCCGCGACCTTGAC GATGCC	64	This project
Vibrio cholerae cpn60	GCCACCGTACTGGCTC AAGCGATT	AACCGCTGCCACTTTCACG ATGCC	56	This project
Arcobacter sp. *cpn60*	GCAACTGTTTTAGCTC ACTCTATT	AACAGCAGCAATATGTAA TGAACC	48	This project
Campylobacter jejuni cpn60	GCAACGGTTTTAGCTC ATGCAATT	CACTGCTGAAATATTAAG AACACC	50	This project
Yersinia enterocolitica cpn60	GCAACTGTATTGGCTC AATCCATT	AACAGCGGCAACTTTAAC AATACC	50	This project

ARRAY #1

	A1	B1	C1	D1
1	*Campylobacter*-70-cpn1	*Campylobacter*-70-cpn2	*Campylobacter*-70-16S1	*Campylobacter*-70-16S2
2	*Legionella*-70-cpn1	*Legionella*-70-cpn2	*Legionella*-70-16S1	*Legionella*-70-16S2
3	*Pseudomonas*-70-cpn1	*Pseudomonas*-70-cpn2	*Pseudomonas*-70-cpn3	*Pseudomonas*-70-16S
4	*Vibrio*-70-cpn1	*Vibrio*-70-cpn2	*Vibrio*-70-cpn3	*Vibrio*-70-16S
5	*Yersinia*-70-cpn1	**50% DMSO**	*Yersinia*-70-16S1	*Yersinia*-70-16S2
6	*Aeromonas*-70-cpn1	*Aeromonas*-70-cpn2	*Aeromonas*-70-16S1	**50% DMSO**
7	*E.coli*-70-cpn1	*E.coli*-70-cpn2	*E.coli*-70-16S1	*E.coli*- 70-16S2
8	*Arcobacter*-70-cpn1	*Arcobacter*-70-cpn2	*Arcobacter*-70-16S1	*Arcobacter*-70-16S2
9	*Arabidopsis* G4S-50	GFP-50	**50% DMSO**	*Arcobacter*-70-16S3

ARRAY #2

	A2	B2	C2	D2
1	*Campylobacter*-amp-cpn	*Legionella*-amp-cpn	*Campylobacter*-amp-16S	*Legionella*-amp-16S
2	*Pseudomonas*-amp-cpn	*Aeromonas*-amp-cpn	**50% DMSO**	**50% DMSO**
3	*Vibrio*-amp-cpn	*Arcobacter*-amp-cpn	*Vibrio*-amp-16S	Arcobacter-amp-16S
4	*E.coli*-amp-cpn	*Yersinia*-amp-cpn	*E.coli*-amp-16S	*Yersinia*-amp-16S
5	*Salmonella*-amp-cpn	**50% DMSO**	*Salmonella*-amp-16S	*Shigella*-amp-16S
6	*Arabidopsis* G4S-50	GFP-amp	*E.coli*-amp-cpn*	*E.coli*-amp-16S*
7	**50% DMSO**	**50% DMSO**	**50% DMSO**	**50% DMSO**
8	**50% DMSO**	**50% DMSO**	**50% DMSO**	**50% DMSO**
9	**50% DMSO**	**50% DMSO**	**50% DMSO**	**50% DMSO**

cpn60 70-mer oligonucleotides

16S 70-mer oligonucleotides

controls

cpn60 amplicons

16S amplicons

Figure 4-3. Probe Pattern of Microarray Prototype 2.0.

4.2.2 Amplicon Generation and Labeling Procedure

PCR amplification was done using 1 µL of pure bacterial culture lysates. Two sets of universal primers were used for PCR amplification of 16S rDNA (F1 GAG-TTT-GAT-CCT-GGC-TCA-G and R2 G(T/A)A-TTA-CCG-CGG-C(T/G)G-CTG) and *cpn60* genes (H279 GAI-III-GCI-GGI-GA(T/C)-GGI-ACI-ACI-AC and H280 (T/C)(T/G)I-(T/C) (T/G)I-TCI-CC(A/G)-AAI-CCI-GGI-GC(T/C)-TT) with high content in inosine for *cpn60* primers. The annealing temperatures were 58°C for 16S rDNA and 43°C for *cpn60* gene amplifications. Purified amplicons were labeled with the Mirus' Cy5 *Label* IT Nucleic Acid Labeling kit according to the manufacturer's instructions and incubated three hours at 37°C in the dark. The unreacted labeling reagents were removed from the labeled nucleic acid using a QIAquick PCR purification kit (Qiagen).

4.2.3 Microarray Hybridization

Microarrays were prehybridized at 37°C for 1 h under a coverslip (22x22 mm; Sigma) in 15 µl of prewarmed solution A (DIG Easy Hyb buffer [Roche] containing 10 µg of tRNA and 10 µg of denatured salmon sperm DNA). After the coverslip was removed by dipping the slide in 0.1X SSC (1X SSC contained 150 mM NaCl and 15 mM trisodium citrate [pH 7]), the array was rinsed briefly in water and dried by centrifugation at room temperature in 50 mL conical tubes for 5 min at 150 g. Fluorescently labeled DNA was chemically denatured as described by the manufacturer and added to 15 µL of a fresh solution of prewarmed solution A. Hybridization was carried out overnight 4 h or 1 h at different temperatures to determine the best hybridization conditions. After hybridization, the coverslip was then removed and placed in 0.1X SSC, and the microarray was washed once in prewarmed 1X SSC-0.1% SDS solution, once in prewarmed 0.1X SSC-0.1% SDS solution and once in 0.1X SSC for 10 min at 42°C. After the array was dried by centrifugation (150 x *g*, 5 min, room temperature), the array was analyzed with a fluorescence scanner (Canberra-Packard, Mississauga, Ontario, Canada).

4.2.4 Results: Microarray Prototype 2.0 Detection Limit

To determine specificity and sensitivity of the prototype 2.0, different amounts of targeted amplicons were hybridized on the two sub-arrays present on the slide. To avoid the risk of sample preparation bias, the same preparation was used for the two sub-arrays. The results obtained for oligonucleotide sub-array are presented in Figure 4-4 whereas those obtained with amplicon sub-array are presented in Figure 4-5.

The results obtained with the 70-mer oligonucleotide sub-array showed that, even if 0.1 ng can be weakly detected, 1 ng of a specific amplicon is the lowest acceptable detection limit of this type of microarray. No cross-hybridization was observed for *cpn60* probes. Four cross-hybridizations were observed with 16S rDNA probes. Although there are some cross-hybridizations with *Aeromonas, Yersinia,* and *Arcobacter* probes, only one out of two or three of the designated oligonucleotides gave a positive signal. This demonstrates that one oligonucleotide per target is not enough and can lead to false positive results. The combination of multiple specific oligonucleotides allows one to validate the results as well as to detect false-positive hybridizations.

Hybridization with the *E. coli* amplicon probes gave some cross-hybridization with the *cpn60* from *Yersinia* because of their sequence homology; a strong cross-hybridization signal was also seen with the 16S amplicon probe for *Vibrio* and a weaker cross-hybridization signal with the *Arcobacter* 16S probe.

The amplicon sub-array was more sensitive and allowed the detection of 0.1 ng of specific target. Nevertheless, more cross-hybridizations were observed when using the amplicon sub-array compared to the 70-mer oligonucleotide sub-array, especially for 16S rDNA probes.

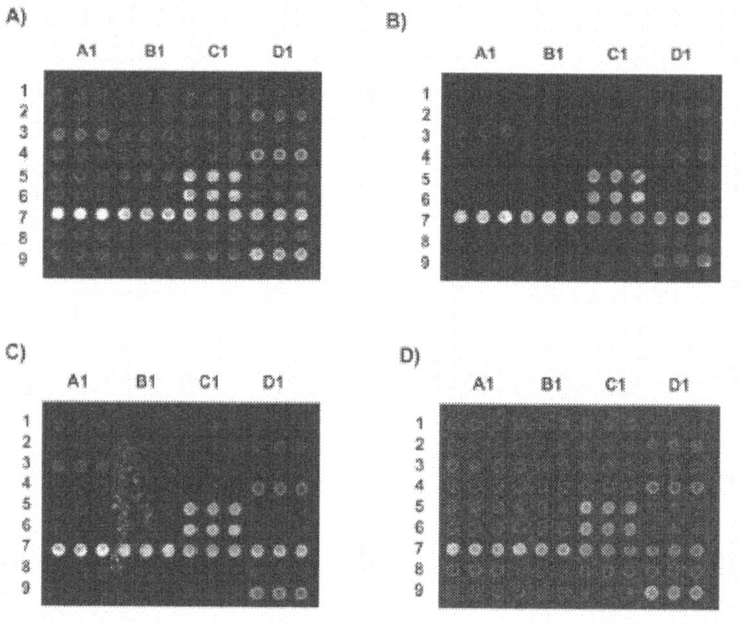

Figure 4-4. Hybridization of Different Concentrations of *E. coli cpn60* and 16S rDNA Amplicons on 70-mer Oligonucleotide Sub-arrays of Prototype 2.0 A) 10 ng of *cpn60* and 10 ng of 16S rDNA Cy5-labeled Amplicons, B) 5 ng of *cpn60* and 5 ng of 16S rDNA Cy5-labeled Amplicons, C) 1 ng of *cpn60* and 1 ng of 16S rDNA Cy5-labeled Amplicons, D) 0.1 ng of *cpn60* and 0.1 ng of 16S rDNA Cy5-labeled Amplicons.

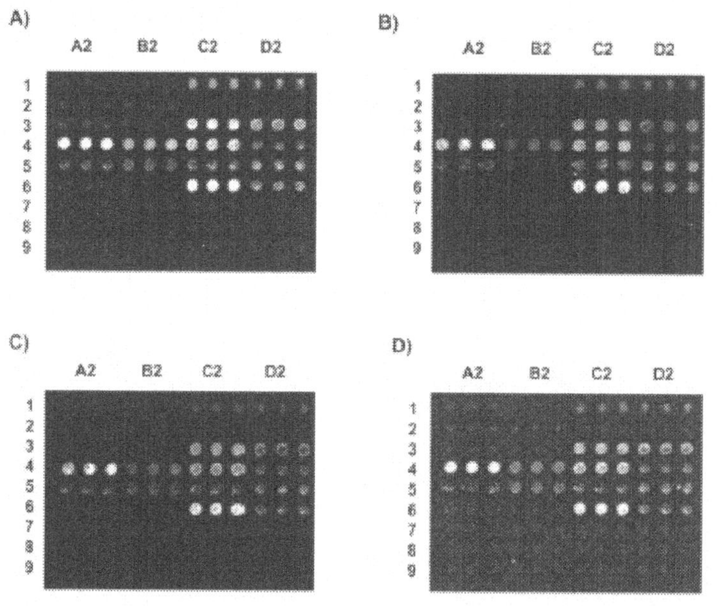

Figure 4-5. Hybridization of Different Concentrations of *E. coli cpn60* and 16S rDNA Amplicons on Amplicon Sub-arrays of Prototype 2.0 A) 10 ng of *cpn60* and 10 ng of 16S rDNA Cy5-labeled Amplicons, B) 5 ng of *cpn60* and 5 ng of 16S rDNA Cy5-labeled Amplicons, C) 1 ng of *cpn60* and 1 ng of 16S rDNA Cy5-labeled Amplicons, D) 0.1 ng of *cpn60* and 0.1 ng of 16S rDNA Cy5-labeled Amplicons.

4.2.5 Conclusions

- In general, amplicon probes appeared to be more sensitive than oligonucleotide probes.
- Amplicons probes for 16S DNA were less specific and generated more cross-hybridizations than oligonucleotide probes for 16S rDNA discrimination due to their length and to the presence of highly conserved areas in the sequences.
- Oligonucleotide probes appeared to be specific and sensitive enough to be used with amplicon targets.

4.3 16S rDNA, *cpn60*, and *wecE* Amplicons on 18- to 26-mer Oligonucleotide Arrays with Short Amino-Linkers

The results obtained using prototype 2.0 demonstrated that a combination of 70-mer oligonucleotide 16S rDNA and *cpn60* probes allows a better discrimination of targets in comparison to the amplicons probes. However, cross-hybridization was also observed with the 70-mer probes. Therefore, to bypass this specificity limitation, we designed prototype 3.0 with short (18-26 mers) oligonucleotides probes. Nevertheless, in the case of Enterobacteriaceae, 16S and *cpn60* sequences were close enough to generate cross-hybridization reactions even with the shorter oligonucleotides and consequently, they may be difficult to discriminate on the basis of these sequences. A specific common feature of this family is the synthesis of enterobacterial common antigen (ECA). The *wec* gene cluster is implicated in the synthesis of ECA and represents a good target for the broad-spectrum detection of Enterobacteriaceae. This gene cluster has previously been shown to allow the use of universal primers yet to also allow the design of very specific probes for Enterobacteriaceae (*101*). Ten different *wecE* short oligonucleotide probes (approximately 20 bases each) were designed and printed on the prototype 3.0 to enhance its discriminating power.

4.3.1 Design

The nucleotide sequences of all the 18- to 26-mer oligonucleotide sequences are reported in Table 4-5. We compared the printing of the same oligonucleotides with and without 5' end amino linkers. Aminolinkers are amine-reactive groups coupled to a chain composed of six carbons (C6). This C6 chain links the amino reactive groups on the glass slide surface to the oligonucleotide probes. The aminolinker role is to link the oligo to the surface of the slide covalently and to orient the immobilized DNA in a more readily available position during the hybridization. The pattern of microarray prototype 3.0 is presented in Figure 4-6.

Table 4-5. Oligonucleotides Used for Prototype 3. All Oligonucleotides Were Produced with and without C6 Amino Linkers at Their 5' End.

Oligonucleotide	Sequence (5'-3')	size (mer)	Tm (°C)
CM_COLI_462	GAG CAA AGG TAT TAA CTT TAC TCC C	25	55.1
KL-16S-Ecoli	CGT CAA TGA GCA AAG GTA TTA AC	23	53.6
KL-cpn60-Ecoli	TTC TTC AAC TGC AGC GGT AAC A	22	58.6
KL-16S-Yer1	CAC AAA GGT TAT TAA CCT TTA TGC C	25	54.3
KL-16S-Yer2	CGC CGG CAA AGT AGT AAA CTA C	22	57.1
KL-cpn60-Yer1	GTG CAA TCA GCT CAC CCA CAG	21	59.7
KL-cpn60-Yer2	AAG TTC AAT AGA ACC GGT TTC TG	23	54.7
KL-cpn60-Salmonella	TTC AAC CGC CGC AGC AAC CGC T	22	68.3
KL-16S-Salmonella	TTG CTG CGG TTA TTA ACC ACA ACA	24	59.4
KL-16S-Vibrio	GTA TTA GCC ATC GTT TCC AAT G	22	53.1
KL-cpn60-Vib1	GAC GAG AGG CTT TCG CTA CGC	21	61.2
KL-cpn60-Vib2	GCC CAC GCT AGA GTC TGA GTT AGC	24	62.3
EUB338	GCT GCC TCC CGT AGG AGT	18	66.7
KL-wecE-E.coli	TGG GAA CCA TTG GTC ATA TTG G	22	56.5
KL-wecE-Salmonella	TAG GGA CGA TCG GTC ACA TCG G	22	61.5
KL-wecE-Y.enterocolitica	CAC TAT CGG CCA TAT TGG TTG CT	23	58.9
KL-wecE-Y.pestis	TCT GGG TAC CAT TGG CCA TAT TG	22	58.5
KL-wecE-Klebsiella	ATC GCA CGC TGG TGG AAC GC	22	64.3
KL-wecE-Citrobacter	AAA ATT AGT CGA ACG TGC AGA GAT	24	56.6
KL-wecE-Enterobacter	TCG TGC GCT GGT TGA GCG TGC	21	66.5
KL-wecE-Serratia	ATC CGG CCC TGA TCG ATC GG	20	62.3
KL-wecE-Proteus	CAT CTA TTT TTA CCC TTA TGA AGG GC	26	55.6
KL-wecE-Providencia	CGC TTT ATA GGA CGC GAA TTG TAC	24	57.5
KL-araG4s1	AAC AAC AAC ATC TGG CGT AAG AGT G	25	59.4
KL-araG4s2	CCA AAC AAT GCT TGG CCA GCC CAC C	25	66.9
CM_RHO_211	GGC TCA TCC TGC ACC AGT AAA C	22	59.4

	A	B	C	D	E	F
1	KL-C6-cpn-Ecoli	KL-C6-16S-Ecoli	CM_C6_ECOLI_462	KL-cpn60-Ecoli	KL-16S-Ecoli	CM_COLI_462
2	KL-C6-cpn-Yer1	KL-C6-16S-Yer2	KL-C6-16S-Yer1	KL-cpn60-Yer1	KL-16S-Yer2	KL-16S-Yer1
3	KL-C6-16S-Salmonella	KL-C6-cpn-Salmonella	KL-C6-cpn-Yer2	KL-16S-Salmonella	KL-cpn60-Salmonella	KL-cpn60-Yer2
4	KL-C6-cpn-Vib2	KL-C6-cpn-Vib1	KL-C6-16S-Vibrio	KL-cpn60-Vib2	KL-cpn60-Vib1	KL-16S-Vibrio
5	KL-C6-wecE-E.coli	***buffer***	EUB338-C6	KL-wecE-E.coli	***buffer***	EUB338
6	KL-C6-wecE-Y.pestis	KL-C6-wecE-Y.enterocolitica	KL-C6-wecE-Salmonella	KL-wecE-Y.pestis	KL-wecE-Y.enterocolitica	KL-wecE-Salmonella
7	KL-C6-wecE-Enterobacter	KL-C6-wecE-Citrobacter	KL-C6-wecE-Klebsiella	KL-wecE-Enterobacter	KL-wecE-Citrobacter	KL-wecE-Klebsiella
8	KL-C6-wecE-Providencia	KL-C6-wecE-Proteus	KL-C6-wecE-Serratia	KL-wecE-Providencia	KL-wecE-Proteus	KL-wecE-Serratia
9	CM_C6_RHO_211	KL-C6-araG4s2	KL-C6-araG4s1	CM_RHO_211	KL-araG4s2	KL-araG4s1
10	KL-C6-cpn-Ecoli	KL-C6-16S-Ecoli	CM_C6_ECOLI_462	KL-cpn60-Ecoli	KL-16S-Ecoli	CM_COLI_462
11	KL-C6-cpn-Yer1	KL-C6-16S-Yer2	KL-C6-16S-Yer1	KL-cpn60-Yer1	KL-16S-Yer2	KL-16S-Yer1
12	KL-C6-16S-Salmonella	KL-C6-cpn-Salmonella	KL-C6-cpn-Yer2	KL-16S-Salmonella	KL-cpn60-Salmonella	KL-cpn60-Yer2
13	KL-C6-cpn-Vib2	KL-C6-cpn-Vib1	KL-C6-16S-Vibrio	KL-cpn60-Vib2	KL-cpn60-Vib1	KL-16S-Vibrio
14	KL-C6-wecE-E.coli	***buffer***	EUB338-C6	KL-wecE-E.coli	***buffer***	EUB338
15	KL-C6-wecE-Y.pestis	KL-C6-wecE-Y.enterocolitica	KL-C6-wecE-Salmonella	KL-wecE-Y.pestis	KL-wecE-Y.enterocolitica	KL-wecE-Salmonella
16	KL-C6-wecE-Enterobacter	KL-C6-wecE-Citrobacter	KL-C6-wecE-Klebsiella	KL-wecE-Enterobacter	KL-wecE-Citrobacter	KL-wecE-Klebsiella
17	KL-C6-wecE-Providencia	KL-C6-wecE-Proteus	KL-C6-wecE-Serratia	KL-wecE-Providencia	KL-wecE-Proteus	KL-wecE-Serratia
18	CM_C6_RHO_211	KL-C6-araG4s2	KL-C6-araG4s1	CM_RHO_211	KL-araG4s2	KL-araG4s1

C6-oligonucleotides in phosphate buffer

C6-oligonucleotides in 50% DMSO

C0-oligonucleotides in phosphate buffer

C0-oligonucleotides in 50% DMSO

Figure 4-6. Probe Pattern of the Microarray Prototype 3.0. Each Area Represents the Position of a Specific Probe on the Chip.

4.3.2 Amplicon Generation and Labeling Procedure

PCR amplification was done using 1 µL of pure bacterial culture lysates. Three sets of universal primers were used for PCR amplifications. The sequence and the annealing temperature used for these PCRs are presented in Table **4-6**. Purified amplicons were labeled with the Mirus' Cy5 *Label* IT Nucleic Acid Labeling kit according to the manufacturer's instructions and incubated three hours at 37°C in the dark. The unreacted labeling reagents were removed from the labeled nucleic acid using a QIAquick PCR purification kit (Qiagen).

Table 4-6. Primer Pair Sequences, Amplicon Sizes, and Primer Annealing Temperatures. K=G or T; Y=C or T; R=A or G; W=A or T; N= A, T, C or G; I= inosine.

Genes	Forward (5'-3')	Reverse (5'-3')	Size (bp)	Annealing Temperature (°C)
16S rDNA	F1: GAGTTTGATCCTGGCTCAG	R2: G(TA)ATTACCGCGGC(TG)GCTG	528	57
cpn60	WDF: GAIIIIGCIGGIGAYGGNACNACNAC	WDR: KIYKITCICCRAANCCNGGNGCYTT	555	43
wecE	wecE1: AGGGCGTGATGTCCACTTAC	wecE2: GAAGAACTGGCTGCGGTTAG	188	55

4.3.3 Microarray Hybridization

In this part of the work, hybridization patterns of a mix of 16S and *wecE* amplicons (25 ng each Cy5 labeled-amplicon), and a mix of *cpn60* and *wecE* amplicons (25 ng of each Cy5 labeled-amplicon) were compared for several bacterial strains. The specificity of the oligonucleotide probes and the utility of the aminolinkers were tested.

4.4 Conclusions

- Short 18 to 26-mer oligonucleotide probes were used and no cross-hybridization reaction was observed.
- The presence of C6 linkers had no real advantage compared to conventional printing. In fact, because many cross-hybridization reactions were observed, the specificity of the C6 amino-linked oligonucleotides appears somewhat lower in comparison to oligonucleotides without amino linker.
- The *wecE* probes are specific to the different species-specific *wecE* genes. Therefore, this gene can be used to discriminate among different Enterobacteriaceae species.

4.5 Final Prototype: 16S rDNA, *cpn60*, and *wecE* Amplicons on 18- to 26-mer Oligonucleotide Arrays

4.5.1 Design

The prototype 4.0 contains the same combination of short 18- to 26-mer oligonucleotide probes specific for 16S rDNA, *cpn60* and *wecE* printed on microarray prototype 3.0 except that the oligos were printed exclusively in dimethylsulfoxide (DMSO) buffer on Corning GAPS II slides. The oligonucleotide probe sequences are shown in Table 4-7.

Table 4-7. Oligonucleotides Printed on Prototype 4.0.

Oligo name	Sequences	Species	Array position
KL-16S-Salmonella	TTG CTG CGG TTA TTA ACC ACA ACA	*Salmonella* sp.	E1
KL-16S-Ecoli	CGT CAA TGA GCA AAG GTA TTA AC	*E. coli* , *Shigella* sp.	F1
CM_ECOLI_462	GAG CAA AGG TAT TAA CTT TAC TCC C	*E. coli* , *Shigella* sp.	G1
KL-16S-Yer1	CAC AAA GGT TAT TAA CCT TTA TGC C	*Y.enterocolitica*	H1
KL-16S-Yer2	CGC CGG CAA AGT AGT AAA CTA C	*Yersinia* sp.	E2
KL-16S-Vibrio	GTA TTA GCC ATC GTT TCC AAT G	*Vibrio* sp.	F2
RB_Salm_wayc	AGC CTG ATG CAG CCG TAT GCA G	*Salmonella waycross*	G2
RB_Salm_hout	TAC CGC AGC TAA TTG ACG TTA C	*S. houten, S. bareilly*	H2
RB_Salm_parA	ATC ACA TGT GCC CAG ATG CCA T	*S. paratyphi A*	E3
RB_Salm_parC	CTT TCC AGA GAT GAG TTT GTG C	*S. paratyphi C*	F3
RB_Salm_heid	ACA TCC ACG GGA AGT TTT CAG A	*S. heidenberg*	G3
RB_Salm_agona	GAA GCA GCT TGC ACG TAG CTG A	*S. agona*	H3
RB_Salm_enteritidis	GAT CCA TTT GTG CCT TAG GGA A	*S. enteritidis*	E4
RB_Salm_thompson C1	GGT TAA GTC CCG CTA CGA TCG C	*S. thompson C1*	F4
EUB338	GCT GCC TCC CGT AGG AGT	Eubacterium	G4
BUFFER			H4
KL-wecE-Proteus	CAT CTA TTT TTA CCC TTA TGA AGG GC	*Proteus* sp.	E5
KL-araG4s1	AAC AAC AAC ATC TGG CGT AAG AGT G	*Arabidopsis thaliana*	F5
CM_RHO_211	GGC TCA TCC TGC ACC AGT AAA C	*Rhodococcus coprophilus*	G5
KL-araG4s2	CCA AAC AAT GCT TGG CCA GCC CAC C	Arabidopsis thaliana	H5
KL-cpn-Ecoli	TTC TTC AAC TGC AGC GGT AAC A	*E. coli O157: H7*	A1
KL-cpn-Yer1	GTG CAA TCA GCT CAC CCA CAG	*Y.enterocolitica*	B1
KL-cpn-Yer2	AAG TTC AAT AGA ACC GGT TTC TG	*Y.enterocolitica*	C1
KL-cpn-Vib1	GAC GAG AGG CTT TCG CTA CGC	*Vibrio cholerae*	D1
KL-cpn-Vib2	GCC CAC GCT AGA GTC TGA GTT AGC	*V. cholerae, V. mimicus*	A2
KL-cpn-Salmonella	TTC AAC CGC CGC AGC AAC CGC T	*S. typhimurium*	B2
PB_Salm_univ	TCG TAC GCT AAT TGA ACG CGC G	*Salmonella* sp.	C2
KL-wecE-Salmonella	TAG GGA CGA TCG GTC ACA TCG G	*S. typhimurium*	D2
KL-wecE-Ecoli	TGG GAA CCA TTG GTC ATA TTG G	*E. coli* , *Shigella* sp.	A3
KL-wecE-Klebsiella	ATC GCA CGC TGG TGG AAC GC	*Klebsiella*	B3
KL-wecE-Serratia	ATC CGG CCC TGA TCG ATC GG	*Serratia*	C3
PB_Yers_enter1_modifie	CAC TAT CGG CCA TAT TGG TTG CTT T	*Yersinia* sp.	D3
PB_Yers_enter2	GCT TAT CAA TGA CCC GTC ACT GA	*Yersinia* sp.	A4
KL-wecE-Y.pestis	TCT GGG TAC CAT TGG CCA TAT TG	*Yersinia pestis*	B4
KL-wecE-Y.enterocolitica	CAC TAT CGG CCA TAT TGG TTG CT	*Yersinia* sp.	C4
KL-wecE-Enterobacter	TCG TGC GCT GGT TGA GCG TGC	*Enterobacter* sp.	D4
PB_Ente_univ	TTT AGC TTC CAT GAA ACC AAA	*Enterobacteriaceae*	A5
PB_Ente_univ2	AGC TTC CAT GAA ACC AAA AA	*Enterobacteriaceae*	B5
KL-wecE-Providencia	CGC TTT ATA GGA CGC GAA TTG TAC	*Providencia*	C5
KL-wecE-Citrobacter	AAA ATT AGT CGA ACG TGC AGA GAT	*Citrobacter*	D5

▒: *wecE* probes ▒: control probes
☐: 16S rDNA probes ■: *cpn60* probes

	A	B	C	D	E	F	G	H
1	*E. coli*	*Y. enterocolitica*	[illegible]	*Vibrio cholerae*	*Salmonella* sp.	*E. coli, Shigella* sp.	*E. coli, Shigella* sp.	*Y. enterocolitica*
2	*V. cholerae, V. mimicus*	*S. enterica*	*Salmonella* sp.	*S. typhimurium*	*Yersinia* sp.	*Vibrio* sp.	*Salmonella waycross*	*S. houten, S. bareilly*
3	*E. coli, Shigella* sp.	*Klebsiella* sp.	*Serratia* sp.	*Yersinia* sp.	*S. paratyphi* A	*S. paratyphi* C	*S. heidelberg*	*S. agona*
4	*Yersinia* sp.	*Yersinia pestis*	*Yersinia* sp.	*Enterobacter* sp.	*S. enteritidis*	*S. thompson* C1	EUB338	BUFFER
5	*Enterobacteriaceae*	*Enterobacteriaceae*	*Providencia* sp.	*Citrobacter* sp.	*Proteus* sp.	*Arabidopsis thaliana*	*Rhodococcus coprophilus*	*Arabidopsis thaliana*

▒: *wecE* probes ▒: control probes

☐: 16S rDNA probes ■: *cpn60* probes

Figure 4-7. Probe Pattern of Microarray Prototype 4.0. Each Area Represents the Position of a Specific Probe on the Chip.

4.5.2 Amplicon Generation and Labeling Procedure

PCR amplifications were performed as described in the standard PCR procedure (Appendix E). Three primer sets were used for the PCR reactions (Table **4-6**). The amplified products were purified using the QIAquick PCR purification kit (Qiagen) according to the manufacturer's instructions.

After purification, amplicons were labeled with the Mirus' Cy5 *Label* IT Nucleic Acid Labeling kit and the unreacted labeling reagents were removed from labeled nucleic acids by using the QIAquick PCR purification kit (Qiagen) as described in Standard labeling procedure (Appendix F).

4.5.3 Hybridization Procedure

The hybridization procedure used is described in the standard microarray hybridization procedure (Appendix G). Different amounts of labeled amplicons were used depending on whether they were derived from pure bacterial culture DNA or a complex DNA mixture.

4.5.4 Results: Microarray Prototype 4.0 Detection Limit

The detection limit of the microarray prototype 4.0 was determined. This work was performed using amplicons generated from genomic DNA extracted from pure cultures of *E. coli, S. typhimurium,* and *Y. enterocolitica* and from an environmental wastewater sample (complex DNA mixture). Combinations of different DNA were tested to evaluate detection limits and specificity:

1. Determination of the detection limit of three PCR reactions (16S, *cpn60,* and *wecE*) using pure bacterial culture DNA;
2. Determination of the prototype 4.0 specificity using fragments amplified from pure bacterial culture DNA;
3. Determination of the prototype 4.0 detection limit using fragments amplified from a binary mixture of DNA;
4. Determination of the prototype 4.0 detection limit using fragments amplified from a tertiary mixture of DNA;

5. Determination of the prototype 4.0 detection limit using fragments amplified from a complex mixture of DNA (*Salmonella* DNA and wastewater-extracted DNA).

4.5.4.1 Detection Limit of the 16S, *cpn60,* and *wecE* PCR Fragments

PCR fragments were generated using different *S. typhimurium* genomic DNA concentrations (ranging between 50 ng to 50 fg) as templates. The DNA quantity added in each PCR reaction is estimated on the basis that 50 ng of genomic DNA correspond to 10^7 genomes.

Similar detection limits were obtained for the 16S and *wecE* PCR reactions. For both, the minimal DNA quantity that can be detected is 500 fg, corresponding to approximately 100 bacterial genomes. The sensitivity of these two reactions is high and appropriate for the detection of bacteria present in low quantity in complex mixtures. Concerning the *cpn60* PCR reaction, the detection limit is 5 pg, representing 1,000 genomes. This threshold is acceptable but some bacteria present in low quantities in complex mixtures may not be detected by the *cpn60* PCR amplification. Amplicon specificity was verified on prototype 4.0 and the hybridization patterns obtained demonstrated their specificity for all the *Salmonella* probes (data not shown).

4.5.4.2 Determination of Prototype 4.0 Specificity Using Fragments Amplified from Pure Bacterial Culture DNA

Twenty-five ng of *E. coli*, *S. typhimurium,* or *Y. enterocolitica* 16S amplicons was separately hybridized on prototype 4.0. The anticipated hybridization patterns were observed with the hybridization of *E. coli* and *S. typhimurium* 16S amplicons (Figure 4-8). However, only one of the two *Yersinia*-specific probe gave a signal when *Y. enterocolitica* 16S amplicons were hybridized on the microarray. It was also noted that the oligonucleotide designed to detect the 16S rDNA for *Salmonella* serotype *waycross* hybridized indiscriminately to all amplicons tested. A sequence analysis was performed between the oligonucleotide sequence and the *E. coli*/*Yersinia* 16S sequences. This analysis demonstrated that for *E. coli* and *Y. enterocolitica*, approximately 10 consecutive bases could potentially hybridize with this oligonucleotide. This similarity is apparently sufficient for DNA annealing when 37°C is used as the hybridization temperature.

The anticipated hybridization patterns were also obtained with each of the bacterial *cpn60* and *wecE* fragments hybridized on the microarray (Figure 4-9 and Figure 4-10). No cross-hybridization was observed.

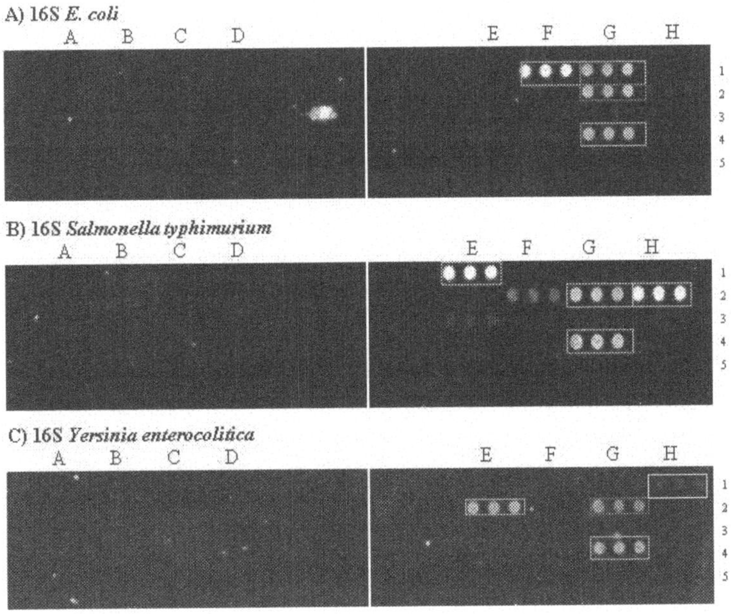

Figure 4-8. Hybridization of 16S rDNA Amplicons on Prototype 4.0. Hybridization of A) 25 ng of *E. coli* 16S Amplicons. B) 25 ng of *S. typhimurium* 16S Amplicons. C) 25 ng of *Y. enterocolitica* 16S Amplicons.

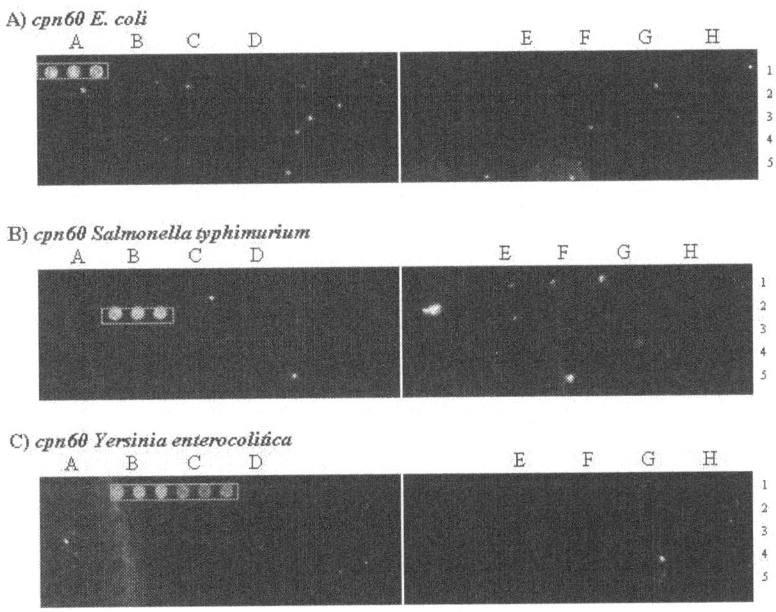

Figure 4-9. Hybridization of *cpn60* Amplicons on Prototype 4.0. Hybridization of A) 25 ng of *E. coli cpn60* Amplicons. B) 25 ng of *S. typhimurium cpn60* Amplicons. C) 25 ng of *Y. enterocolitica cpn60* Amplicons.

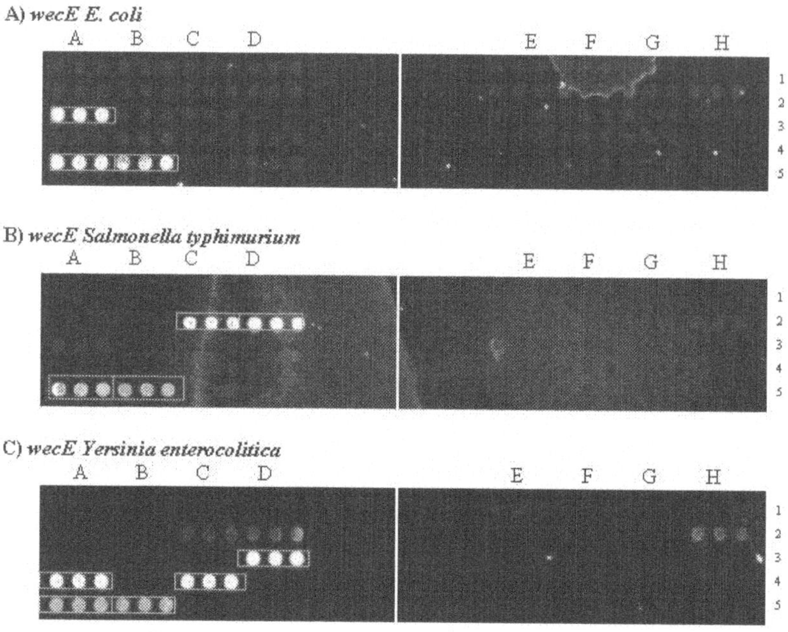

Figure 4-10. Hybridization of *wecE* Amplicons on Prototype 4.0. Hybridization of A) 25 ng of *E. coli wecE* Amplicons. B) 25 ng of *S. typhimurium wecE* Amplicons. C) 25 ng of *Y. enterocolitica wecE* Amplicons.

4.5.4.3 Determination of the Prototype 4.0 Detection Limit Using Fragments Amplified from a Binary Mixture of DNA

To test the limits of our method, it was crucial to determine the lowest quantity of *Salmonella* DNA detectable by hybridization on prototype 4.0. This aim was achieved by using a binary mixture of genomic DNA. The majority of the mixture was composed of genomic DNA extracted from *E. coli* to mimic the conditions encountered in the wastewater sample where *E. coli* organisms are predominant over *Salmonella* or *Yersinia*. To the *E. coli* genomic DNA, different quantities of *Salmonella* DNA were added. The concentrations of each DNA added in the mixture are presented in Table 4-8. An amount of 1 μg of labeled amplicons obtained after each PCR reaction was hybridized on prototype 4.0 to determine the detection threshold of the microarray.

After hybridization of each of the 16S, *cpn60* and *wecE* amplicons on prototype 4.0, a *Salmonella*-specific signal was detected when 1 μg of a 16S, *cpn60,* or *wecE* PCR fragment pool coming from a 0.1% or 1% *Salmonella* binary mixture was hybridized. Mixtures containing less than 0.1% *Salmonella* DNA resulted in no signal (data not shown). Therefore, the prototype 4.0 detection limit is 1 μg of amplicon (either 16S, *cpn60,* or *wecE*) provided from a binary mix containing 0.1% (50 pg) *Salmonella* genomic DNA.

Table 4-8. Quantities of *Salmonella* and *E. coli* Genomic DNA Added in the Different Binary Mixtures.

***Salmonella* DNA (ng)**	***E. coli* DNA (ng)**	**% *Salmonella***
0.5	49.5	1
5×10^{-2}	50	0.1
5×10^{-3}	50	0.01
5×10^{-4}	50	0.001

4.5.4.4 Determination of the Prototype 4.0 Detection Threshold Using Fragments Amplified from a Ternary Mixture of DNA

We found that *Salmonella* can be detected from a binary DNA mixture containing 0.1% *Salmonella* (50 pg *Salmonella* DNA and 50 ng *E. coli* DNA). To determine if the presence of genomic DNA from other microorganisms could affect this detection limit, ternary mixtures were generated. The different ternary mixtures used as PCR templates are presented in Table 4-9. One μg of labeled amplicons obtained after each PCR reaction was hybridized on prototype 4.0 to determine the detection threshold of the microarray.

Salmonella specific hybridization signals were detected with fragments amplified from any ternary DNA mixtures containing 0.1% *Salmonella* (data not shown). The signal intensities were similar, no matter if the *Yersinia* genomic DNA ratios added in the mix corresponded to 0.1 or 1.0%. Therefore, the *Yersinia* genomic DNA presence does not affect the detection limit determined using binary mixtures.

Table 4-9. *Salmonella, E. coli,* and *Yersinia* Genomic DNA Quantities Added in the Different Ternary Mixtures.

***Salmonella* DNA (ng)**	***Yersinia* DNA (ng)**	***E. coli* DNA (ng)**	**% *Salmonella***	**% *Yersinia***
5×10^{-2}	0.5	49.5	0.1	1
5×10^{-2}	5×10^{-2}	50	0.1	0.1

4.5.4.5 Determination of the Prototype 4.0 Detection Threshold Using Fragments Amplified from a Complex Mixture of DNA

The detection limit for bacterial DNA contained in a complex environmental wastewater sample might be different from the threshold determined with bacterial DNA extracted from pure culture due to different factors such as mixture complexity, DNA integrity, etc. In order to determine if detection thresholds are similar, different *S. typhimurium* genomic DNA ratios were spiked into wastewater extracted DNA sample #73. The different *Salmonella* DNA/wastewater DNA mixtures are reported in Table 4-10.

Table 4-10. *Salmonella* Genomic DNA Quantities Added in the Environmental Complex Extracted DNA.

***Salmonella* DNA (ng)**	**Wastewater Sample #73 DNA (ng)**	**% *Salmonella***
2.5	47.5	5.0
0.5	49.5	1.0
5×10^{-2}	50	0.1
0	50	0

The 16S and *cpn60* hybridization results demonstrated that for the wastewater sample, without the addition of *Salmonella* genomic DNA, no signal was observed for the *Salmonella*-specific probes.

As presented in Figure 4-11 and Figure 4-12, 16S and *cpn60 Salmonella* specific hybridization signals were detected with fragments amplified from environmental DNA mixtures containing 0.1% *Salmonella*. Thus, the *Salmonella* detection threshold in an environmental DNA mixture is 10^4 genomes and is identical to the one obtained using amplicons generated from binary or ternary mixtures. However, this environmental detection limit is only an estimate because the quantity of *Salmonella* potentially present in wastewater sample #73 was not known.

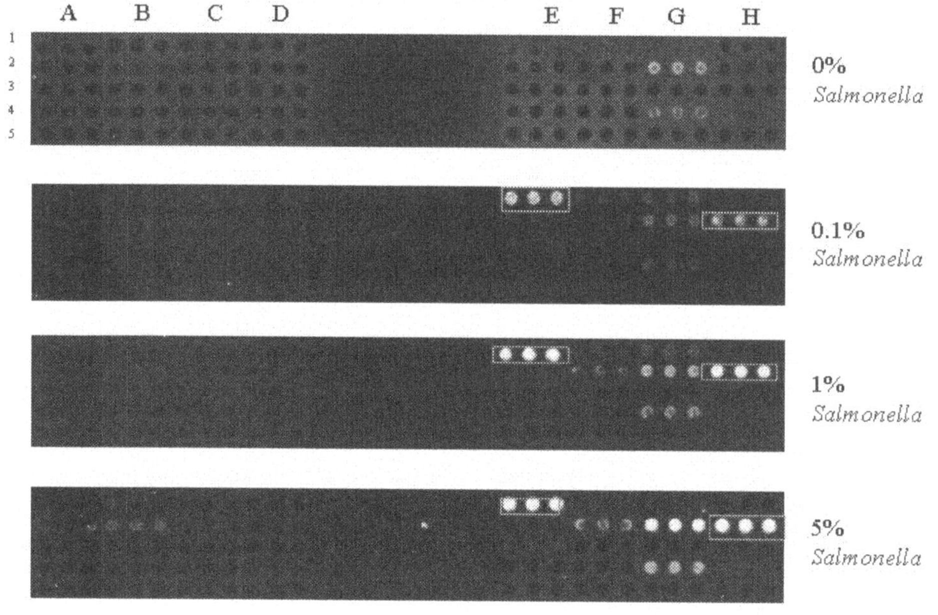

Figure 4-11. Hybridization Results of 1 µg of 16S rDNA Amplicons Derived from Different *Salmonella* DNA Ratios Spiked in Wastewater Extracted DNA. 0%, 0.1%, 1.0%, and 5.0% of *Salmonella* DNA Were Added.

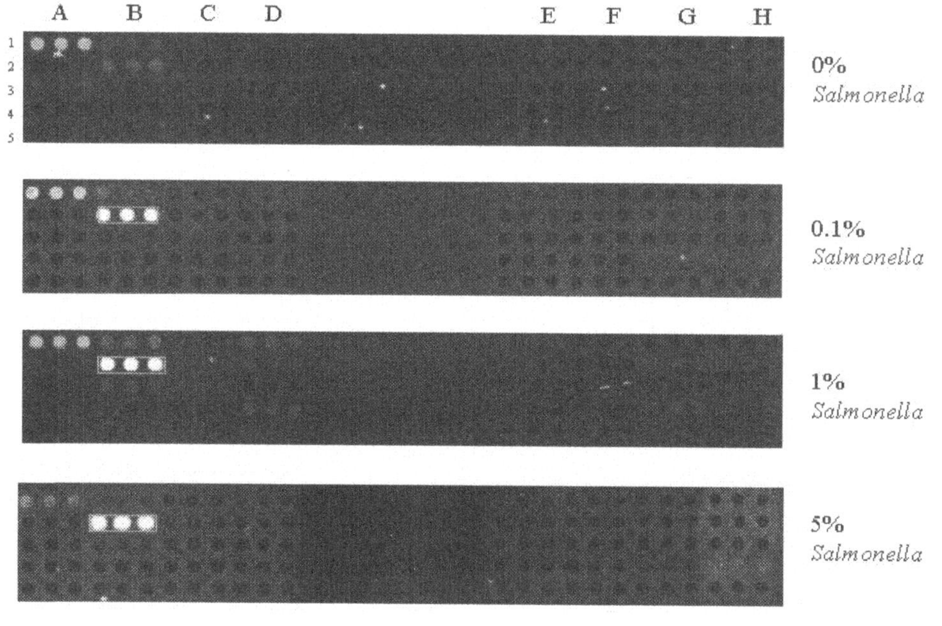

Figure 4-12. Hybridization Results of 1 µg of *cpn60* Amplicons Derived from Different *Salmonella* DNA Ratios Spiked in Wastewater Extracted DNA. 0%, 0.1%, 1.0%, and 5.0% of *Salmonella* DNA Were Added.

In contrast to the 16S and *cpn60* fragments, the *wecE* hybridization results demonstrated that *Salmonella*-specific signals were detected using the *wecE* genes for wastewater sample #73 without the addition of *Salmonella* genomic DNA. However, because *Salmonella* can be detected using the *wecE* fragments, the detection limit appears to be lower for the *wecE* procedure than for the 16S and *cpn60* procedure. This can be explained by the fact that the *wecE* PCR reaction is more specific than the 16S or *cpn60* reactions. Thus, the *wecE* PCR reaction allows the amplification of the *wecE* genes present only in the genome of the *Enterobacteriaceae* a much smaller subset of species-specific genes in contrast to the 16S or *cpn60* reactions that allow the amplification of fragments from the majority of the microorganisms present in the wastewater sample.

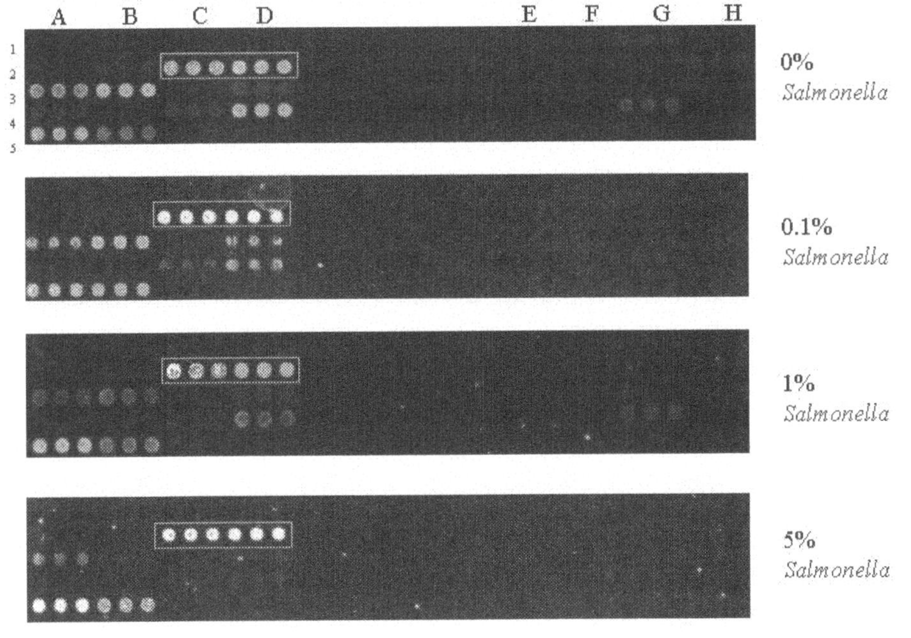

Figure 4-13. Hybridization Results of 1 μg of *wecE* Amplicons Derived from Different *Salmonella* DNA Ratios Spiked in Wastewater Extracted DNA. 0%, 0.1%, 1.0% and 5.0% of *Salmonella* DNA Were Added.

4.5.5 Conclusions

The 16S, *cpn60* and *wecE* probes are specific to the three microorganisms tested (*E. coli*, *S. typhimurium* and *Y. enterocolitica*). The only cross-hybridization reaction observed is with the *Salmonella* serotype *waycross* probe that hybridized with the *E. coli*, *S. typhimurium,* and *Y. enterocolitica* 16S amplicons. It may due to the high sequence homology between oligonucleotide probe and these bacterial 16S rDNA.

Similar PCR detection limits were determined for the 16S and *wecE* reactions. For both reactions, the minimal DNA quantity that can be detected is 500 fg corresponding to approximately 100 bacterial genomes in pure culture. The *cpn60* PCR detection limit is higher than the two reactions described above. This PCR detection limit corresponds to 5 pg, which represents 1,000 genomes in pure culture.

Using binary DNA mixtures, the prototype 4.0 detection threshold was determined and corresponded to 0.1% of *Salmonella* DNA. This detection limit was not affected by the addition of a third DNA component in the binary mixture.

The *Salmonella* detection threshold in an environmental DNA mixture is 0.1% and is identical to the one obtained using amplicons generated from binary or ternary mixtures. However, this detection limit is only an estimate because the *Salmonella* quantity was not determined prior to the amplification.

No specific hybridization signal was observed for *Salmonella* when the 16S and *cpn60* fragments amplified from extracted wastewater DNA (without *Salmonella* added) were hybridized. In contrast, *Salmonella*-specific signals were detected in the same wastewater sample when the *wecE* fragments were hybridized. Thus, the detection limit for the *wecE* amplicons appears to be lower than the 16S and *cpn60* genes. Since *wecE* PCR targets consensus and conserved Enterobacteriaceae-specific sequences instead of consensus bacterial sequences in the taxonomic16S or *cpn60* assays, it detects all members of the *Salmonella* family in contrast to the latter techniques. Moreover, detection of *wecE* amplicons is accomplished by specific probes and the level of sensitivity is better than 16S or *cpn60* for detecting these enteric pathogens. Therefore, to increase the sensitivity of the prototype for pathogens present in wastewater, it might be advantageous to use amplicons generated from PCR reactions specific for a limited group of bacteria (e.g., species-, family-, genus-specific) instead of universal PCR reaction specific for a broader spectrum of bacteria (e.g., 16S, *cpn60*).

Please Note: The Tables and Figures found in Chapters 4.0 and 5.0 can be viewed in color in the Product Catalog section of the WERF Website at www.werf.org.

CHAPTER 5.0

DEVELOPMENT AND EVALUATION OF MICROARRAYS FOR BACTERIAL SOURCE TRACKING

Bacterial source tracking (BST) was developed to determine the sources of fecal bacteria from environmental samples (e.g., from human, livestock, wildlife origins). This methodology is also been called fecal sourcing or fecal typing in the literature. At present, BST can reliably determine if fecal bacteria are from human or animal sources. It is unknown at this time if BST can eventually achieve distinctions between different types of livestock (e.g. cattle, horse, swine, poultry, etc.) or wildlife (e.g., deer, waterfowl, raccoon, etc.) or pets (dogs, cats, etc.). BST methods have been used primarily with *E. coli* and the fecal streptococci. However, these methods have also been applied to bifidobacteria, *Bacteroides-Prevotella* group, and coliphages though to a much lesser degree. The first objective of this chapter is therefore to present, *via* a short bibliographic review, the indicator organisms that can be used to determine the origin of the fecal pollutants found in water. Those indicator organisms were subsequently used to design the Bacterial Source Tracking microarray.

In general, normal animal and human microflora are composed of diverse bacteria and in different concentrations (Table 5-1). It is possible to determine some species that are over-represented among animals when compared to humans or vice versa. For example, *Bacteroides*, *Bifidobacillus* and *Streptococcus* bacteria are found in both sources but some specific species are only found among one of them (*102, 103*). In addition, some bacteria are found in both animals and humans but their concentrations are different. For example, *Eubacterium* sp. and *Fusobacterium* sp. are more abundant in human feces than in animal feces (*102-105*). In contrast, *Selenomonas* sp. and *Megasphaera* sp. are more abundant in animal feces than in human feces (*106*).

The analysis of the literature on normal animal and human microflora illustrates the difficulties in confirming that the presence of a specific bacterium indicates the source of the contamination. Therefore, a selection of indicators to determine the general features of the contamination source profile seemed to be the best way to explore these distinctions. It is the reason why, in addition to the selected bacterial indicators, the detection of fecal indicators such as *E. coli* or *Clostridium perfringens* was considered useful.

Table 5-1. Different Frequencies and Concentrations of Bacteria Found in Normal Animal and Human Microflora.

Organism	Positive human sample frequency (%)	Concentration in human fecal sample (colonies per gram)	Positive animal sample frequency (%)	Concentration in animal fecal sample (colonies per gram)	References
Bacteroides					
distasonis	52,5	10^{10}	<5	-	(*102*)
thetaiotaomicron	86,5	10^{10}	<5 (rabbit)	-	(*102*)
uniformis	90	10^{8}	(pig) rare	$<10^{5}$	(*103*)
vulgatus	69,5	10^{10}	None	-	(*102*)
Bifidobacterium					
Adolescentis	54,6	10^{10}	None	-	(*102*)
animalis	<5	-	95	$>10^{9}$	
breve	9,9	10^{10}	None	-	(*102*)
infantis	24,8	10^{10}	<5 (calf)	-	(*102*)
longum	21,3	10^{10}	<5 (calf)	-	(*102*)
thermophilum	none	-	28	10^{6}	(*103*)
Butyrivibrio					
fibrisolvens	none	-	90 (pig and ruminal)	$>10^{9}$	(*107*)
Enterobacter					
cloacae	16,3	10^{7}	None	-	(*102*)
Enterococcus					
casseliflavus	rare	low	>75	$>10^{9}$	(*106*)
cecorum	none	-	>90	$>10^{9}$	(*108*)
Eubacterium					
aerofaciens	48,9	10^{10}	rare (chicken)	-	(*102, 104*)
contorium	26,2	10^{10}	None	-	(*102*)
cylindroides	24,3	10^{10}	None	-	(*102*)
Facklamia					
sp.	rare	low	90	10^{9}	(*109*)
Fusobacterium					
prausnitzii	90	10^{9}	40	10^{5}	(*103, 105*)
Megasphaera					
elsdenii	rare	-	>90	10^{8}	(*106*)
Pediococcus					
urinaeequi	rare	-	90	$<10^{5}$	(*109*)
Pseudomonas					
aeruginosa	10,6	10^{5}	Rare	-	(*102*)
Selenomonas					
ruminantium	rare	low	20 (pig)	10^{5}	(*106*)
Salinicoccus					
sp.	none	-	>90	10^{10}	(*106*)
Streptococcus					
alactolyticus	none	-	>50 (chicken)	10^{9}	(*105, 108*)
Intestinalis	none	-	>90	10^{10}	(*103, 110*)
lactis	28,4	10^{8}	Rare	low	(*102*)

5.1.1 Design

The most logical target gene for designing a BST microarray is the 16S rDNA sequence. Those sequences, coding for the small subunit of ribosomal RNA, are now the most widely used informational macromolecule for bacterial systematic studies at the family, genus, species, and subspecies levels. The 16S rDNA gene contains conserved sequences that can be used to infer natural relationships between distantly related species and variable regions that can be used to separate closely related ones. Such a 16S rDNA sequence-based identification technique will substantially facilitate the ecological study and the control of difficult-to-culture microorganisms. The ready availability of these sequences facilitates the creation of molecular taxonomic tools.

The protoype BST microarray design was created on the basis of 19 bacterial genera. A total of 78 oligonucleotide probes based on bacterial 16S rDNA sequences were chosen which represent 25 fecal animal indicators, 25 fecal human indicators, 10 Enterobacteriaceae bacteria, 15 human and animal fecal indicators, and three positive controls (*Eubacterium oligo* and Enterobacteriaceae genera oligos). They represent a summary of the literature and laboratory personnel investigation on 16S rDNA genes probes specific for BST.

Once the candidate bacterial species as a potential fecal indicator had been selected, their 16S rDNA sequences were regrouped in a local sequence bank and specific 16S rDNA oligonucleotides were designed using the OligoPicker software (version 2.3.1. for Linux, available from http://pga.mgh.harvard.edu/oligopicker/index.html) (*100*). To increase the hybridization specificity for our BST prototype, 22-mer oligonucleotides were designed over the length of the total 16S rDNA gene (position 4 to 1468). Sequence comparisons were submitted to the National Center for Biotechnology Information (Bethesda, MD) for BLAST searches in GenBank. Multiple DNA alignments were performed by using the CLUSTALW program (*111*). Oligonucleotide probes were synthesized with a Commercial Nucleic Acid Synthesizer, Model ABI394 (Integrated DNA Technologies, Coralville, IA, USA). The probe sequences, their size as well as the corresponding bacterial indicator and their reference are listed in Table 5-2.

Table 5-2. List of Oligonucleotides Used for the Primary BST Microarray Design.

Strain	Oligo probe (5' to 3')	Length of sequence	Position	Tm	G+C content	Accession number	References
Bacteroides							
acidofaciens	TGG CCG CTT ACA CTA TAT CGC A	22	16S rDNA 845-824	59.8	50	AB021159	This study
acidofaciens	CTG TAC TCA AGG ATG CCA GTA T	22	16S rDNA 647-626	55.3	45.5	AB021157	This study
ovatus	GCC CCA CGA CTT GTA TGT GTT A	22	16S rDNA 56-35	58.5	50	AB050108	This study
sp. ASF519	CCA GAG CTG TCA ATA TGC GTT C	22	16S rDNA 988-967	57.3	50	AF157056	This study
sp.	GCG TTT CGG CCT CTG TCA ATT G	22	16S rDNA 1100-990	60.4	54.5	AF319778	This study
sp.	GTT AAG CCT GTA	21	16S	55.9	47.6	X83953	This study

	GCT AGC GTT		rDNA 24-4				
sp. *(human)*	CGG ACA TGT GAA CTC ATG AT	20	16S rDNA 175-156	53.5	45	AF294909	This study
thetaiotaomicron	AGC GAG TAG TCA TTG TTT ACT G	22	16S rDNA 818-797	53.8	40.9	AB050109	This study
thetaiotaomicron	CGA AAG GCT ATC CCC GAG TTA T	22	16S rDNA (VR1-3) 170-149	57.6	50	M58763	(*112*)
thetaiotaomicron	TGC AAG CAA ACT GAA ATG CTG C	22	16S rDNA 96-75	59.1	45.5	L16486	This study
Bifidobacterium							
adolescentis	CTC CCA GTC AAA AGC GGT T	19	16S rDNA 442-434	56.3	52.6	M58729	(*113*)
adolescentis	GAC ATG CAT CCA ACT GGA G	19	16S rDNA 202-184	54.1	52.6	M58729	(*114*)
adolescentis	AAG GAC ATG CAT CCA ACT GGA G	22	16S rDNA 205-184	58.7	50	M58729	(*115*)
adolescentis	TCT CAC GGC CGG AGC AAG CTC C	22	16S rDNA 72-51	67	68.2	AF275881	This study
aerophilum	CAC ACC ATG CAG CAT GAT GGA G	22	16S rDNA 176-155	59.9	54.5	AY174107	This study
breve	GTA GGA TTA GTC CCA ATC AGG A	22	16S rDNA 1477-1456	54.5	45.5	AJ311605	This study
breve	AGC CCG ATG GAT ACC CGT TCG A	22	16S rDNA 66-45	63.7	59.1	AJ311605	This study
gallinarum	TCC CAT GCG GTC ATC TGG AGC A	22	16S rDNA 210-189	64.6	59.1	D86191	This study
infantis	ATC AGG ATT TCA CCT TAG ACG G	22	16S rDNA 1468-1447	55.5	45.5	AJ311604	This study
inopinatum	ACC GAA TGG TCC ACT GGG CTG C	22	16S rDNA 186-165	65.5	63.6	D89332	This study
longum	TCC CAA CAC ACC ATG CGG TGT G	22	16S rDNA 191-172	63.8	59.1	AJ311606	This study
longum	CTC TAC GAC CGT CGG GAA C	19	16S rDNA 991-973	57.5	63.2	U10152	(*113*)
magnum	AGC GCA AAT CCG CCG TTA AGC G	22	16S rDNA 616-595	64.5	59.1	D86193	This study
minimum	CCC ATG AGG TCA TGT GGA GCA T	22	16S rDNA 200-179	60.6	54.5	M58741	This study
pseudolongum	CTT TCC CAC ACT CGC ATG CGC T	22	16S rDNA 200-179	63.9	59.1	D86195	This study

Psychro-aerophilum	GAC CCC ATC TCA TAC CGC TAA C	22	16S rDNA 210-189	58.1	54.5	AY174108	This study
sp.	CAT CCG GCA TTA CCA CCC	18	16S rDNA 165-148	56.3	61.1	D89379	(*116*)
sp.	TAA GCG ATG GAC TTT CAC ACC	21	16S rDNA 586-566	56.2	47.6	D89379	(*116*)
sp. oral strain	CCC CTA GCA AGC TAG AAA GGA T	22	16S rDNA 65-44	57.5	50	AF385524	This study
Butyrivibrio							
fibrisolvens	TTC AAG ATT CTG CCG AAG CTT C	22	16S rDNA 69-48	57	45.5	U41172	This study
Clostridium							
perfringens	GTT GAA TGA TGA TGC CAT CTT T	22	16S rDNA 198-177	53.2	36.4	Y12669	(*117*)
ramnosum	CGC CAC TCG AGC ACA AGT GCT C	22	16S rDNA 92-71	63.9	63.6	X73440	This study
Enterobacter							
cloacae	TGG CAC CTG AAG GTC AGT CTT T	22	16S rDNA 730-709	60.1	50	AF511434	This study
Enterobac-teriaceae							
ENT1	CCG CTT GCT CTC GCG AG	17	16S rDNA 1269-1253	59	70.6	AY098487	(*118*)
Sonde D	TCT CGC GAG GTC GCT TCT CTT	21	16S rDNA 1251-1241	61.5	57.1	AY098487	(*119*)
Enterococcus							
avium	TCA AAC CGA AAC CAT GCG GTT TC	23	16S rDNA 187-165	60.2	47.8	Y18274	(*120*)
casseliflavus/ gallunarum/ flavescens	GCG GAA AAT AGT GTT ATA CGG	21	16S rDNA 197-177	52.4	42.9	AF367978	(*120*)
casseliflavus	AGC CTC AGC GTT CCA GTT ACA G	22	16S rDNA 774-753	60.5	54.5	AJ301826	This study
cecorum	ATG TTT CTA TGC CGG TTT CAG G	22	16S rDNA 1440-1461	56.9	45.5	AJ301827	This study
cecorum	CTC ATT TTC TTC CGG TGT AGC AAG	24	16S rDNA 103-80	57.6	45.8	AJ301827	(*120*)
faecalis	CCC TCT GAT GGG TAG GTT	18	16S rDNA 110-93	53.6	55.6	AF477496	(*120*)
faecalis	GTA GGA GTT TTG CGG CCG TGT C	22	16S rDNA 355-334	61.8	59.1	AJ271855	This study
faecalis	CTC GCG GTC TAA CGA CTC GTT G	22	16S rDNA 1245-	60.8	59.1	AF477496	This study

			1224				
faecium	CCC GTG TCT CTA GTC CAA TGT G	22	16S rDNA 346-325	58.2	54.5	AB012213	This study
saccharolyticus	CCT CTT TTC CAA TGA ACC GAA G	22	16S rDNA 99-78	55.3	45.5	AJ301839	This study
sp.	GGC TTT ACC CCC TTT GAT GGG T	22	16S rDNA 137-116	61	54.5	AF349929	This study
sp.	GCA GTC TTG GCT AGG AGT GCC C	22	16S rDNA 1179-1158	63.2	63.6	AB015233	This study
sp.	GTT ATC CCC TGC TGA TGG GCA G	22	16S rDNA 151-130	61.1	59.1	AJ420805	This study
Escherichia							
coli	GAG CAA AGG TAT TAA CTT TAC TCC C	25	16S rDNA 462-438	55.1	40	Z83204	(*121*)
coli	TTC CCG AAG GCA CAT TCT	18	16S rDNA 1021-1004	54.8	50	Z83204	(*121*)
Eubacteria							
EUB338	GCT GCC TCC CGT AGG AGT	18	16S rDNA	66.7	59.5		(*122*)
Eubacterium							
aerofaciens	ACG ACT TAA CCC CCT TAA CCT T	22	16S rDNA 1450-1429	57.3	45.5	AB011816	This study
aerofaciens	GGA GAG GGG TGC CGT TCG ACT T	22	16S rDNA 73-52	64.6	63.6	AB011816	This study
Lactococcus							
lactis	TCC AGT CGG TAC AAG TAC CAA C	22	16S rDNA 69-48	57.5	50	AB100805	This study
Megasphaera							
elsdenii	ACC TTC GAC GGC TGG CTC CAA A	22	16S rDNA 1495-1474	64.9	59.1	M26493	This study
Pseudomonas							
aeruginosa	CAG GAG CAA GCT CCC TTC ATC C	22	16S rDNA 82-61	61	59.1	AB091760	This study
Rhodococcus							
sp.	GGC TCA TCC TGC ACC AGT AAA C	22	16S rDNA 211-190	59.4	54.5	AJ007001	This study
sp.	TGC AAG CCC GCA GTT GAG CTG C	22	16S rDNA 509-488	66.7	63.6	AY233201	This study
Ruminococcus							
palustris	CTA TTC TCT CGA TCG GAA GCA A	22	16S rDNA 88-67	55.4	45.5	AJ296179	This study
Salmonella							
Enterica subsp.	AGA CCA GGT AAC	22	16S	60.2	54.5	U92195	This study

houtenae serovar *Houten*	CTT CTT CGC G		rDNA 990-969				
Enterica serovar *Waycross*	GGC ATG CTG CAT ACG GCT GCA T	22	16S rDNA 409-388	64.5	59.1	U92194	This study
Enterica serovar *Waycross*	GCA TGC TGC ATA CGG CTG CAT C	22	16S rDNA 408-387	62.9	59.1	U92194	This study
paratyphi	ACA AGC TAA TGG CAT CTG GGC A	22	16S rDNA 251-230	60.9	50	AE014489	This study
Enterica serovar *Bovis-morbificans*	ATG GGA AGA GGC CTG GAG GTC C	22	16S rDNA 223-202	64.1	63.6	U92193	This study
Enterica serovar *Chingola*	TCT CGA CTT GAC AGC CGC CTG C	22	16S rDNA 601-580	65.1	63.6	U92192	This study
Selenomonas							
ruminantium	CGA CGG CTA TAT CCT TGC GGT T	22	16S rDNA 1486-1465	60.5	54.5	M62702	This study
Serratia							
Ficaria	TAT TAA GAC ATC GCC CTT CCT C	22	16S rDNA 443-422	55.3	45.5	AJ233428	This study
Shigella							
sonnei	CGA AGG CAC CTT TTC CAT CTC T	22	16S rDNA 1032-1011	58.6	50	X80726	This study
Streptococcus							
bovis	AAC TCT TCC AAC TTT AGC AAG C	22	16S rDNA (V1 region) 95-74	54.9	40.9	AF104113 (animal)	This study
bovis	CCA ACT TTA GCA AGC TAA AGT C	22	16S rDNA (V1 region) 88-67	53.2	40.9	AF104109 (animal)	(*123*)
bovis	CTT CTA CTC TAG CAA GCT AAA GTA G	25	16S rDNA (V1 region) 90-66	53.9	40	AF104114 (human)	(*123*)
gallinaceus	CAG CTA ACA TGT GTC GGC TGC T	22	16S rDNA 177-156	61.2	54.5	AJ307888	This study
gallolyticus	CGC AAC TCC TTC TAC TCT AGC A	22	16S rDNA 90-69	57.4	50	X94337	This study
intestinalis	TCA TCC AAA GGG AGG CAG GCT C	22	16S rDNA 79-58	62.7	59.1	AB002519	This study
intestinalis	GCA GGC TCC CTC TTT CAG GCG T	22	16S rDNA 65-44	65.2	63.6	AB002519	This study
mitis	ACT AAA CCC CGG AAA GGG TCT A	22	16S rDNA 852-831	58.9	50	D38482	This study
mutans	AGT CAG ACA GTT	22	16S	58.1	50	AJ243965	This study

	TCC AGA GCA C		rDNA 639-618				
plutanimalium	AAG AGA TTG GCT TAC TGT CAC C	22	16S rDNA 1277-1256	56	45.5	Y18026	This study

5.1.2 Slide Preparation

All the oligonucleotides were suspended in water to obtain a stock concentration of 100 pmol/µL. A plate was prepared by diluting the oligonucleotides to a final volume of 10 µl in DMSO 50% (v/v) and a final concentration of 50 pmol/µL for all oligonucleotide probes. They were printed in accordance with the design shown in Figure 5-1.

5.1.3 Amplification and Amplicon Labeling Procedures

The 16S rDNA PCR amplification was performed as described in Appendix E. Universal 16S rDNA primer set (F1, 5'- GAG TTT GAT CCT GGC TCA G-3', and R13, 5'- AGA AAG GAG GTG ATC CAG CC-3') was used as described by Dorsch et al. (*98*) to get an amplified product size of 1533 bp. The amplified products were purified using the QIAquick PCR purification kit (Qiagen) according to the manufacturer's instructions.

After purification, the amplicons were labeled with the Mirus' Cy5 *Label* IT Nucleic Acid Labeling kit and the unreacted labeling reagents were removed from labeled nucleic acids by using the QIAquick PCR purification kit (Qiagen) as described in Appendix F.

5.1.4 Hybridization Procedure

The hybridization procedure used is described in Appendix G. Different amounts of labeled amplicons were used depending on whether according if they were amplicons provided from pure bacterial culture DNA or complex DNA mixture.

5.1.5 BST Microarray Prototype Specificity Validation Using Pure Cultures

The BST microarray prototype specificity was determined by the hybridization of 16S rDNA amplicons derived from pure cultures of *Streptococcus salivarus*, *Streptococcus bovis*, *Rhodococcus terrae*, *Rhodococcus equi*, *Escherichia coli*, *Rhodococcus erythropolis*, *Pseudomonas aeruginosa* and *Salmonella paratyphi*, and *Enteroccocus avium*. In most cases, the oligonucleotide probes specificity was demonstrated.

With all tested 16S rDNA amplicons, the designated oligonucleotides gave a positive signal. Only one cross hybridization was observed between *Streptococcus bovis* and *S. salivarus* 16S rDNA amplicons and *Megasphaera elsdenii* 16S rDNA probe (B2). In all hybridizations, the internal control A11 that represents *Arabidopsis thaliana* chlorophyll synthase G4 gave a reproducible and comparable signal.

	A	B	C	D	E	F	G	H	
1	*Enterococcus casseliflavus/ gallinarum/ flavescens*	*Enterococcus avium*	*Clostridium perfringens*	*Butyrivibrio fibrisolvens*	*Bifidobacterium adolescentis*	*Bacteroides thetaiotamicron*	*Bacteroides thetaiotaomicron*	*Bacteroides sp.*	1
2	*Rhodococcus sp.*	*Megasphaera elsdenii*	*Enterococcus cecorum*	*Enterococcus cecorum*	*Bifidobacterium longum*	*Bifidobacterium breve*	*Bifidobacterium adolescentis*	*Bifidobacterium adolescentis*	2
3	*Streptococcus bovis*	*Streptococcus bovis*	*Selemonas ruminantium*	*Rhodococcus sp.*	*Bacteroides sp.*	*Eubacterium aerofaciens*	*Eubacterium aerofaciens*	*Clostridium ramosum*	3
4	*Bifidobacterium magnum*	*Bifidobacterium gallinarum*	*Bifidobacterium aerophilum*	*Streptococcus bovis*	*Bacteroides ovatus*	*Bacteroides acidofaciens*	*Bacteroides acidofaciens*	*Bacteroides sp.*	4
5	*Enterococcus casseliflavus*	*Bifidobacterium psychroaerophilum*	*Bifidobacterium pseudolongum*	*Bifidobacterium minimum*	*Bifidobacterium longum*	*Bifidobacterium inopinatum*	*Bifidobacterium adolescentis*	*Bacteroides thetaiotaomicron*	5
6	*Streptococcus intestinalis*	*Streptococcus gallolyticus*	*Streptococcus gallinaceus*	*Raminococcus palustris*	*Streptococcus mutans*	*Streptococcus mitis*	*Enterobacter cloacae*	*Bifidobacterium sp.*	6
7	*Salmonella enterica*	*Salmonella enterica*	*Enterococcus sp.*	*Streptococcus plutanimalium*	*Bifidobacterium sp.*	*Bifidobacterium sp.*	*Bacteroides sp.*	*Streptococcus salivarius*	7
8	*Salmonella enterica*	*Salmonella enterica*	*Salmonella paratyphi*	*Salmonella enterica*	*Enterobacteriaceae*	*Enterobacteriaceae*	*Bifidobacterium breve*	*Bifidobacterium infantis*	8
9	*Enterococcus saccharolyticus*	*Enterococcus faecium*	*Shigella sonei*	*Serracia ficaria*	Eubacterium	*Escherichia coli*	*Escherichia coli*	*Enterococcus faecalis*	9
10	Buffer	Buffer	Buffer	Buffer	*Enterococcus faecalis*	*Enterococcus faecalis*	*Pseudomonas aeruginosa*	*Lactococcus lactis*	10
11	*Arabidopsis thaliana*	Buffer	Buffer	Buffer	*Enterococcus sp.*	*Enterococcus sp.*	Buffer	Buffer	11

Animal indicator — General indicator

Human indicator — Buffer

Figure 5-1. Probe Pattern of the Bacterial Source Tracking Microarray Prototype Based on Bacterial 16S rDNA.

The results concerning the specificity of the bacterial source tracking microarray prototype demonstrated that this microarray is suitable for the detection of different 16S rDNA amplicons provided from pure cultures. The distinction between closely related bacteria, using 22-mer oligonucleotide probes, is easy to achieve. For example, no cross-hybridization occurs when using *E. coli* and *Salmonella paratyphi* 16S rDNA amplicons or when using different *Streptococcus* species 16S rDNA amplicons, suggesting that the printed oligonucleotide probes are quite specific.

5.1.6 Results: Hybridization with Human Source Wastewater

All the human wastewater samples were collected from the same place but on different days (Table 5-3). The bacterial source tracking analysis by hybridization produced different results among the different samples. The majority of the hybridizations with human wastewater-derived 16S rDNA amplicons gave signals with the *Enterococcus faecium*, *Bacteroides* sp., *B.*

thetaiotaomicron, and Eubacterium probes (Figure 5-2). In some cases, the *Eubacterium aerofaciens*, Enterobacteriaceae, and *Salmonella enterica* probes also gave a signal. Most of the time, a very weak signal was obtain with the *Bifidobacterium adolescentis* and *Bifidobacterium* sp. probes. This suggested that these bacterial species were present at a lower concentration than *Bacteroides,* for example. As demonstrated in Report 6, *E. coli* and other *Enterobacteriaceae* species were probably present but they were not detected by hybridization. These bacteria are universal fecal pollution indicators and were probably also present in animal wastewater samples. The results showed that human fecal indicators were present in these wastewater samples in high concentration. Apparently, none of the animal fecal indicators represented on our BST microarray were present in sufficient concentration to be detected by hybridization.

Table 5-3. Heterotrophic and *E. coli* Colony-Forming Units (CFU) Counts in the Different Human Wastewater Samples. Bacterial Cell Concentrations in Raw Water Were Determined by Culture at 30°C on Plate Count Agar (PCA agar) for Heterotrophic Bacteria Counts and at 44.5°C on mTEC Agar for *E. coli* Counts Following the Methods Described in Appendix C.

Sample	Sampling period	Heterotrophic Cell Concentrations (CFU/ mL)	*E. coli* Concentrations (CFU/ mL)
28	January 15th, 2003	$9.54 \pm 0.48^{*} \times 10^8$	$1.58 \pm 0.16^{*} \times 10^7$
34	February 12th, 2003	$7.62 \pm 0.41 \times 10^8$	$1.62 \pm 0.05 \times 10^7$
48	March 18th, 2003	$1.06 \pm 0.10 \times 10^9$	$1.16 \pm 0.05 \times 10^7$
49	March 18th, 2003	$1.06 \pm 0.10 \times 10^9$	$1.16 \pm 0.05 \times 10^7$
55	April 14th, 2003	$9.10 \pm 0.70 \times 10^8$	$9.50 \pm 1.19 \times 10^6$
61	April 14th, 2003	$9.10 \pm 0.70 \times 10^8$	$9.50 \pm 1.19 \times 10^6$

*Standard deviation

5.1.7 Results: Hybridization with Animal Manure or Soil Runoff after Manure Application

It was extremely difficult to produce 16S PCR fragments with this material presumably because of the presence of inhibitors. The signals obtained after hybridization were weaker than with human samples, even though the same amount of 16S amplicons was hybridized. The internal control, *Arabidopsis thaliana*, showed a weak signal, suggesting poor overall hybridization (data not shown).

It was surprising to observe that most of the samples did not give a signal with the animal indicator probes. The more frequent signals were provided by *Eubacterium*, *Enterococcus faecium*, *Bacteroides* sp., and *Salmonella paratyphi* (Figure 5-3). These bacteria are considered human indicators because they are found in higher concentrations in human intestine but they can be also present in some animal intestines.

The hybridization results obtained with the porcine drainage water samples (#1 and #3) were similar to each other, as were the porcine manure samples (#1 and #2). The manure samples gave more signal than the water samples. The concentration of fecal bacteria is higher in the manure samples than in the water samples (Table 5-4).

Table 5-4. Bacterial Analysis of Animal Contaminated Environmental Samples.

Samples	Sampling Date	Heterotrophic Cell Concentrations	*E. coli* Concentrations (CFU/ mL)
Bovine manure (#1 and #2)	November 5th, 2003	$2.9 \pm 0.5^{*} \times 10^{7}$ CFU/g	$1.3 \pm 0.3 \times 10^{5}$
Water contaminated with bovine manure (#2, #3, #4)	November 7th, 2003	$8.8 \pm 1.4 \times 10^{5}$ CFU/mL	$2.3 \pm 0.1 \times 10^{3}$
Porcine manure (#1, #2, #3)	November 13th, 2003	$1.3 \pm 0.1 \times 10^{7}$ CFU/g	$9.3 \pm 0.1 \times 10^{4}$
Water contaminated with porcine manure (#1, #2, #3)	November 13th, 2003	$7.7 \pm 0.5 \times 10^{6}$ CFU/mL	$3.9 \pm 0.1 \times 10^{4}$

*Standard deviation

5.1.8 Results: Hybridization with Animal Feces

Two DNA extractions of animal feces samples were used to do hybridizations with the BST microarray prototype (Figure 5-4). In these two cases, similar hybridization patterns were obtained, comparable to those obtained from animal source wastewater.

The more frequent signals were produced by *Eubacterium aerofaciens*, *Enterococcus faecium*, *E. faecalis*, and *Bacteroides* sp probes. The *E. coli* specific probe gave a signal in only one case. No specific animal indicator probes gave a signal with these samples. This suggests two possibilities. Firstly, the major part of the fecal microbial population is composed of these bacteria, no matter that it comes from human or animal sources. Using PCR amplification, these predominant microorganisms are amplified and no different BST hybridization pattern between human and animal could be determined. The second possibility is that the specific animal bacterial indicators printed on our BST microarray are not present in the sample.

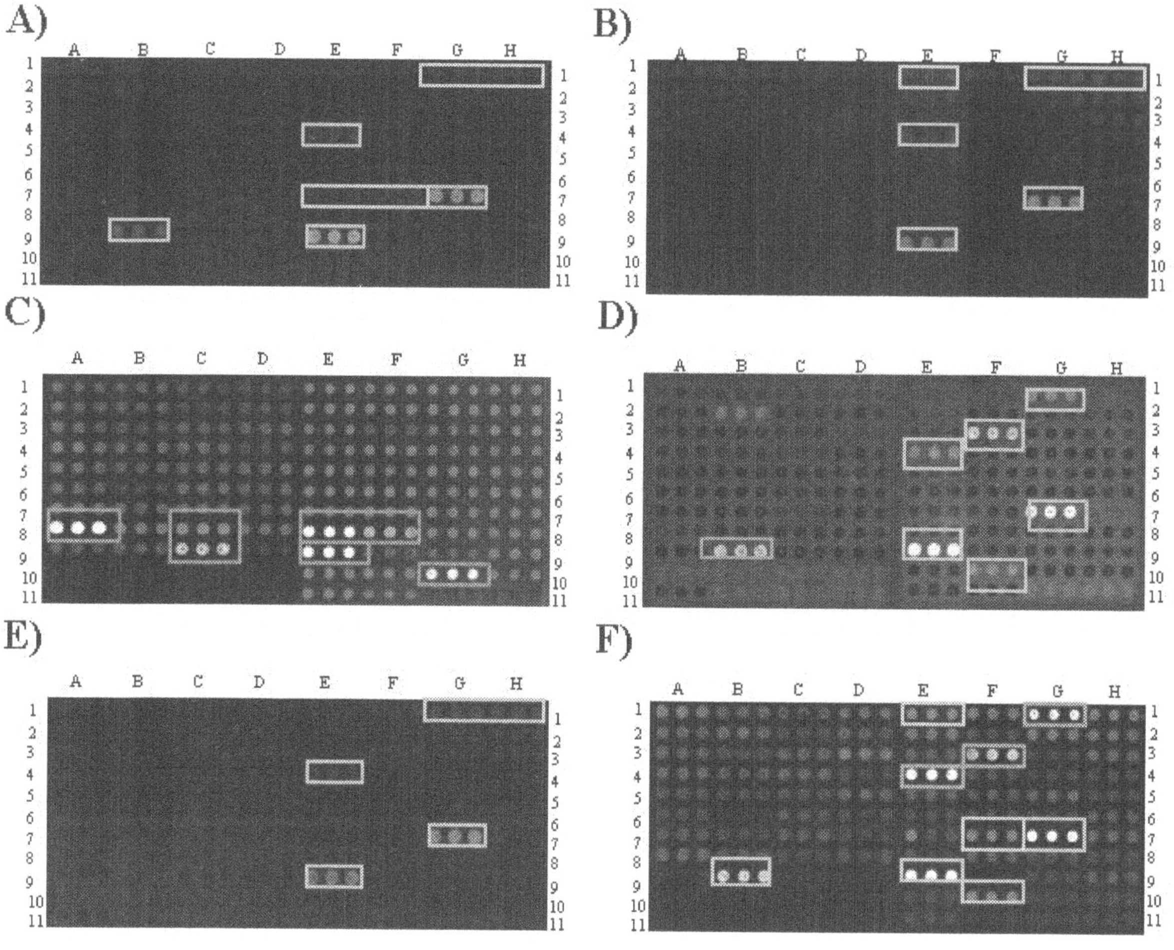

Figure 5-2. Hybridization of 1 µg of 16S rDNA Amplicons Derived from DNA Extracted from Fabreville (Human) Wastewater Samples. A) Sample #28, B) Sample #34, C)Sample #48, D) Sample #49, E) Sample #55, and F) Sample #61. Strong Signals Were Marked by a Green Rectangle and Weak Signals, by a Yellow One.

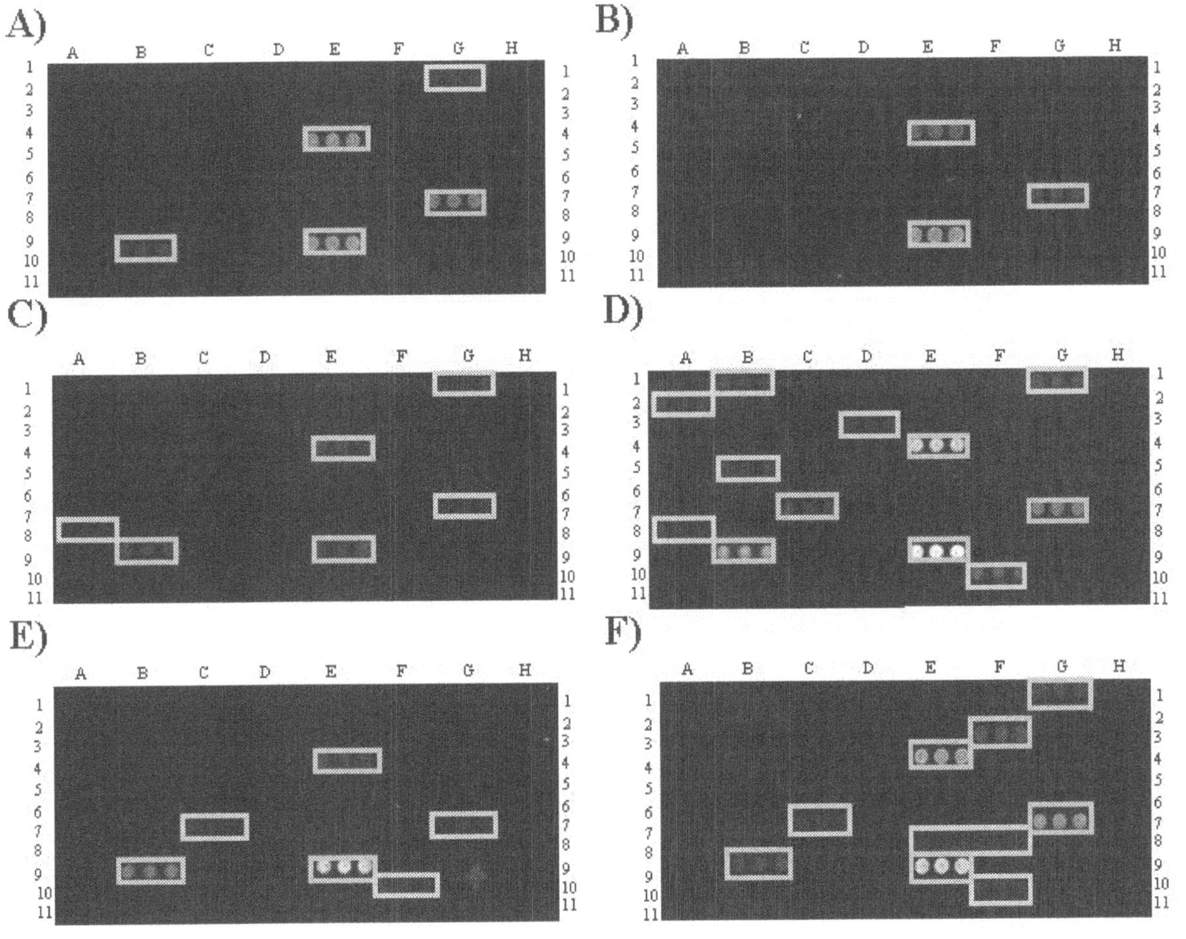

Figure 5-3. Hybridization of 1 μg of 16S rDNA Amplicons Derived from DNA Extraction of Animal Wastewater Samples. A) Porcine Drainage Water Sample #1; B) Porcine Drainage Water Sample #3; C) Porcine manure sample #1; D) Porcine Manure Sample #2; E) Porcine Sample #3; and F) Bovine Drainage Water Samples #2. Strong Signal Was Marked by a Green Rectangle and Weak Signal, by a Yellow One.

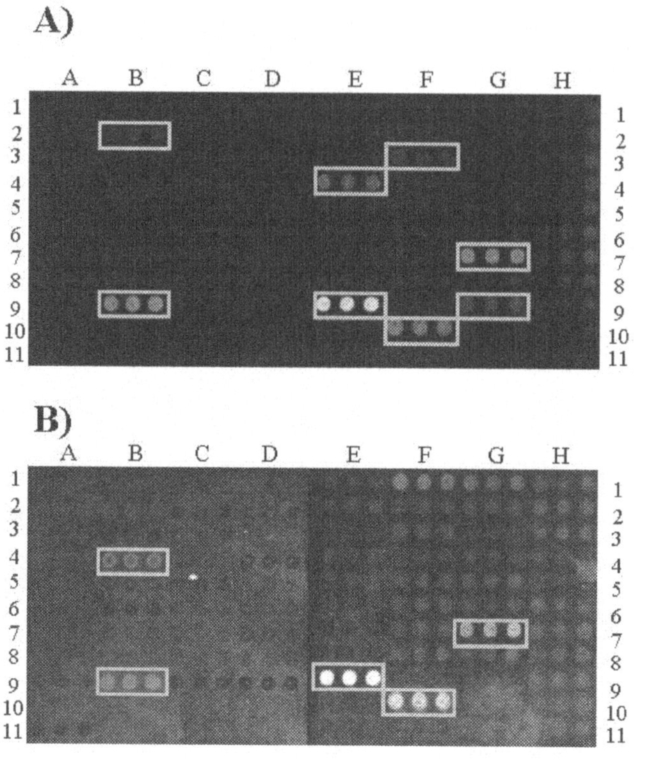

Figure 5-4. Hybridization of 1 μg of 16S rDNA Amplicons Derived from DNA Extraction of Animal Feces Samples. A) Porcine Feces; and B) Goat Feces.

5.1.9 Conclusions

- Oligonucleotides, 22 bases in length, show good specificity and adequate signal strength when hybridized to amplicon targets.
- Specific signals for human fecal indicators have been obtained from wastewater samples from human origin.
- Manure and manure runoff samples only produced signals with general fecal indicators. This suggests that the animal fecal indicators represented on the microarray are either inappropriate for these samples or are present in too low a concentration (perhaps below 0.1%, based on our pathogen detection thresholds).

Please Note: The Tables and Figures found in Chapters 4.0 and 5.0 can be viewed in color in the Product Catalog section of the WERF Website at www.werf.org.

CHAPTER 6.0

CONSIDERATIONS ON THE USABILITY OF DNA MICROARRAYS IN WASTEWATER ANALYSIS

6.1 Time Required for Analysis

With commonly available laboratory equipment, the microarray hybridization procedure requires a timeframe in the range of two and one-half working days. The current need for rapid sample processing to address bioterrorism threats will undoubtedly stimulate development of automated equipment and PCR procedures with faster turnaround times. Such procedures, however, generally target only one organism (such as *Bacillus anthracis*) and may not be applicable to the multiple pathogens of interest in wastewater analysis. The currently developed microarray is expected to find applications in situations where several pathogens of interest are to be detected above the current estimated threshold of 0.1%.

6.2 Estimated Cost of Analysis

Cost of supplies for analysis of one wastewater sample using microarrays is estimated in Table 6-1. Although the cost of approximately $70 appears high, it must be remembered that this microarray provides information on the presence or absence of several pathogens simultaneously. Assuming that 12 pathogens are represented on the array, a cost in the range of $6 per pathogen seems reasonable.

6.3 Reproducibility

Reproducibility of the microarray procedure for detecting wastewater pathogens is fair to good. The main problem areas are the reproducibility of the PCR amplification protocols between samples and the control of the fluorescent background during the hybridization step. Neither of these problems is surprising in light of the complex, "dirty" nature of the wastewater samples (e.g., endogenous fluorescence, PCR inhibitors). The optimized DNA extraction procedure, although helpful, does not totally eliminate such problems.

Table 6-1. Estimation Cost of One Microarray Hybridization Experiment. All Prices Are in U.S. $.

DNA extraction		
DNA extraction from wastewater sample	$3.65/sample	$3.65
Cost for oligonucleotides		
Oligonucleotide 22-mers	$0.41/base	$9.02
Minimum of 40 oligos/ 5,000 slides	($9.02 × 40 oligos)/5,000 slides	$0.07
Cost for one PCR amplification		
Amplification solution (3X)	$0.66/reaction	$1.90
PCR purification	$1.46 /reaction	$1.46
Cost for one microarray hybridization (one assay)		
Slide printing	$23.36/slide	$23.36
Amplicon labeling	$23.36/sample	$23.36
Labeled amplicon purification	$1.46/amplicon	$1.46
Hybridization solutions	$1.46/hybridization	$1.46
Scanning	$14.60/slide	$14.60
Total cost:		**$71.32**

6.4 General Suitability in Wastewater Analysis

The PCR + microarray combination seems destined to remain primarily a research tool at the present time. Its cost, time requirements, and technical complexity do not seem suitable for routine use. Automation of sample collection, sample concentration, DNA extraction, DNA labeling, sample hybridization, and data analysis all seem essential steps before the procedure would become routine. The rapid progress in these areas, driven in part by the bioterrorism threat, will hopefully change this situation within the next few years.

APPENDIX A

STANDARD WASTEWATER SAMPLE CONCENTRATION PROCEDURE

1.0 Objectives/Purpose
To concentrate a raw wastewater sample in small aliquots.

2.0 Safety
Handle wastewater samples with care in biological flow hood.
Wear gloves when handling wastewater samples.
Dispose of contaminated material in biohazardous waste.
Observe Good Lab Practice when handling wastewater suspensions (*124*).

3.0 Materials and Equipment
Raw wastewater sample.
Swinging buck rotor centrifuge (Beckman JS-4.2).
8 L sterile screw cap sampling plastic container (Nalgene)
1 L sterile screw cap plastic centrifugation bottle (Nalgene)
50 mL sterile polypropylene tube with cap for centrifugation.

4.0 Procedures

4.1 Collect approximately 20 L of raw wastewater samples from the effluent of a wastewater treatment plant into three 8 L plastic containers. Store the sample at 4°C until concentration.

4.2 Divide the sample into 900 mL aliquots into 1-L sterile screw cap plastic centrifugation bottles.

4.3 Install the 1 L sterile screw cap plastic centrifugation bottles containing the wastewater aliquots into the buckets of a swinging bucket rotor centrifuge.

4.4 Centrifuge the samples at 3,000 x *g* for 16 minutes at room temperature.

4.5 Discard the supernatant in the contaminated waste container and conserve approximately 300 mL in a separate sterile bottle. Do not disturb the pellet.

4.6 Resuspend each wastewater pellet in the smallest supernatant fraction possible (generally between 6 to 10 mL) using the supernatant conserved in step 4.5.

4.7 Transfer each wastewater aliquot to separate 50 mL sterile polypropylene tubes and store at –20°C until DNA extraction.

APPENDIX B

STANDARD PROCEDURE FOR HETEROTROPHIC BACTERIAL COUNTS ON PCA AGAR

As described in U.S. Environmental Protection Agency (U.S. EPA) (March 2000). Improved enumeration methods for Recreational Water Quality Indicators: Enterococci and *Escherichia coli*. Office of Science and Technology, Washington, D.C. 20460. EPA/821/R-97/004

1.0 Objectives/Purpose
To detect and enumerate the heterotrophic bacteria present in a raw wastewater sample.

2.0 Safety
Handle wastewater samples with care in biological flow hood.
Wear gloves when handling wastewater samples.
Dispose of contaminated material in biohazardous waste.
Observe Good Lab Practice when handling wastewater suspensions (*124*).

3.0 Materials and Equipment
Wastewater samples
Vortex
30±0.5°C incubator
10 mL plastic tubes with cap
9x50 mm Petri dish (Millipore, No. PD-100-4705)

4.0 Procedure

4.1 Homogenize a wastewater aliquot using a vortex mixer.

4.2 Prepare aseptically, in 10 mL sterile plastic tubes with cap, different dilutions of the sample by diluting a wastewater aliquot into 1X phosphate buffer and mix. Select sample volumes based on previous knowledge of the pollution level, to produce 20–80 colonies on the plates (Ex: 10^4, 10^5, 10^6, 10^7 dilutions).

4.3 Uniformly spread 100 μl of the different wastewater dilutions on Plate Count Agar (PCA) plates.

4.4 Incubate the PCA plates at 30±0.5°C in an incubator for 24 hours. After incubation, remove the plate and count aseptically the number of colonies.

4.5 Incubate the PCA plates at 30±0.5°C in an incubator for an additional 24 hours. Remove the plate from the incubator and count the number of colonies observed. The number of colonies counted after 48 hours of incubation should be considered as the heterotrophic bacterial count.

Medium and Buffer Preparation

A. PCA (Plate Count Agar) medium

5.0 g BactoTryptone (Fisher Scientific Co, No. DF0123075)
2.5 g Yeast extract (Difco, No. 212750)
1.0 g glucose (Fisher Scientific Co, No. DB-16B)

Dissolve the mixture in 700 mL of distilled water. Adjust the pH of the solution to 7.2±0.2. Add 15.0 g of Agar (Fisher Scientific Co, No. BP1423-2), bring to 1 L with distilled water. Autoclave at 121°C (15 lb pressure). Aseptically, pour the medium into each Petri dish to a 7-8 mm depth (approximately 25-30 mL), and allow to solidify. Store at 4°C.

B. Phosphate buffer stock solution 800X

Dissolve 34 g of KH_2PO_4 (Laboratoire MAT No. PR-0206) in 500 mL of distilled water. Adjust the pH of the solution to 7.2±0.2, bring to 1 L with distilled water. Autoclave at 121°C (15 lb pressure) for 20 min and store at 4°C.

C. Phosphate buffer working solution 1X

Add 1.25 mL of the 800X phosphate buffer stock solution to 998.75 mL of distilled water. Autoclave at 121°C (15 lb pressure) for 20 min and store at 4°C.

APPENDIX C

STANDARD *ESCHERICHIA COLI* COUNT ON mTEC AGAR PROCEDURE

As described in U.S. Environmental Protection Agency (U.S. EPA) (March 2000). Improved enumeration methods for Recreational Water Quality Indicators: Enterococci and *Escherichia coli*. Office of Science and Technology, Washington, D.C. 20460. EPA/821/R-97/004

1.0 Objectives/Purpose
To detect and enumerate *E. coli* present in a raw wastewater sample.

2.0 Safety
Handle wastewater samples with care in biological flow hood.
Wear gloves when handling wastewater samples.
Dispose of contaminated material in biohazardous waste.
Observe Good Lab Practice when handling wastewater suspensions (*124*).

3.0 Materials and Equipment
Wastewater samples
Membrane filtration system (Millipore Corp)
Vortex
35±0.5°C incubator
44.5±0.2°C incubator
47 mm-diameter, 0.45 μm pore-sized nitrocellulose membrane filters (Millipore Corp, No HAWG04751).
Sterile forceps
10 mL plastic tubes with cap
9x50 mm Petri dish (Millipore, No. PD-100-4705)

4.0 Procedure

4.1 Prepare mTEC Agar and urea substrate as described below.

4.2 Homogenize a wastewater aliquot using a vortex.

4.3 Prepare aseptically, in 10 mL sterile plastic tubes with caps, different dilutions of the sample by diluting a wastewater aliquot into 1X phosphate buffer and mix by pipetting gently. Select sample volumes based on previous knowledge of the pollution level, to produce 20–80 *E. coli* colonies on the membranes (Ex: 1/50, 1/100, 1/500, 1/1000 dilutions). Multiple volumes of the same sample or sample dilutions may be filtered, and the results may be combined.

4.4 Place a sterile membrane filter on the filter base, grid side up, and attach the funnel to the base so that the membrane filter is held between the funnel and the base.

4.5 Filter the sample, and rinse the sides of the funnel at least twice with 20–30 mL of 1X sterile phosphate buffer. Turn off the vacuum, and remove the funnel from the filter base.

4.6 With sterile forceps, remove the membrane filter from the filter base aseptically, and apply it onto the surface of mTEC agar. Avoid the formation of air bubbles between the membrane and the agar surface. If bubbles are observed, re-apply the membrane on the surface of the plate. Apply the forceps around the edge of the filter to fix the filter on the agar properly. Close and invert the dish.

4.7 For all the wastewater dilutions tested, repeat steps 4.4 to 4.6 individually.

4.8 Incubate the mTEC plates containing the membrane on their surface at 35±0.5°C in an inverted position for 2 hours.

4.9 After 2 hours incubation, place the plates at 44.5±0.2°C for 22-24 hours.

4.10 After 22–24 hours incubation, remove the plate from the incubator. Place an absorbent pad in the lid of the same Petri dish and saturate the pad with the urea substrate. With sterile forceps, transfer the membranes from mTEC Agar to the absorbent pads saturated with urea substrate. Incubate at room temperature for 15–20 min.

4.11 After 15–20 min incubation, count the number of yellow, yellow-brown, or yellow-green colonies on the membrane filters. Ideally, select the membrane filter with an acceptable number of 20-80 yellow tinted colonies on the urea substrate. Colonies that are yellow, yellow-brown, or yellow-green are *E. coli*. Colonies that are pink or purple represent nontarget bacteria.

Medium and Buffer Preparation

A. **mTEC Agar (Difco 0334-15-0)**
Ingredients in the commercial preparation:
proteose peptone #3 5.0 g
yeast extract 3.0 g
lactose 10.0 g
NaCl 7.5 g
dipotassium phosphate 3.3 g
monopotassium phosphate 1.0 g
sodium lauryl sulfate 0.2 g
sodium deoxycholate 0.1 g
bromocresol purple 0.08 g
bromophenol red 0.08 g
agar 15.0 g

Preparation: Add 45.3 g of dehydrated mTEC Agar to 800 mL of distilled water in a flask, and heat to boiling until the ingredients dissolve. Adjust pH to 7.3±0.2. Adjust volume to 1 L. Autoclave at 121°C (15 lb pressure) for 20 min, and cool in a 50°C waterbath. Aseptically, pour the medium into each 9x50 mm Petri dish to a 4–5 mm depth (approximately 4–6 mL), and allow to solidify. Store at 4°C.

B. Urea Solution
2.0 g of urea (Difco No. 0190-17)
phenol red 0.01 g (Sigma No. 4633)
Add distilled water to 100 mL final volume. Store at 4°C.

C. Phosphate Buffer Stock Solution 800X
Dissolve 34 g of KH_2PO_4 (Laboratoire MAT No. PR-0206) in 500 mL of distilled water. Adjust the pH of the solution to 7.2±0.2. Add distilled water to a 1 L final volume. Autoclave at 121°C (15 lb pressure) for 20 min and store at 4°C.

D. Phosphate Buffer Working Solution 1X
Add 1.25 mL of the 800X phosphate buffer stock solution to 998.75 mL of distilled water. Autoclave at 121°C (15 lb pressure) for 20 min and store at 4°C.

APPENDIX D

STANDARD WASTEWATER DNA EXTRACTION PROCEDURE

Adaptation of the procedure described in Yu, Z.; Mohn, W. W. Killing two birds with one stone: simultaneous extraction of DNA from activated sludge biomass. *Can. J. Microbiol.* **1999**, *45*, 269-272.

1.0 **Objectives/Purpose**
To extract genomic DNA from human or animal wastewater samples

2.0 **Safety**
Handle wastewater samples with care in biological flow hood.
Wear gloves when handling wastewater samples.
Dispose of contaminated material in biohazardous waste.
Observe Good Lab Practice when handling wastewater suspensions.

3.0 Materials and Equipment
Wastewater sample
Vortex
Centrifuge 16,000 × *g* (Eppendorf ; Angled-centrifuge 5415)
Mini-Beadbeater (Mini-beadbeater-8, BioSpec Products, Model No. 693)
Speedvac (Savant, Model No. SVC200H)
Spectrophotometer
4°C working room
200 µl and 1 mL micropipettes
Glass beads (1 mm in diameter Biospec Inc. No. 11079110) for DNA extraction **from human samples only.**
Glass beads (0.1 mm in diameter Biospec Inc. No. 11079101) for DNA extraction **from animal samples only.**
1.5 mL sterile polypropylene microtubes (Eppendorf)
3.0 mL sterile conical screw cap microtube for bead-beating (Fisher Scientific Co, Cat No. FSSP9700936)

4.0 Procedure

4.1 Homogenize the wastewater sample by vortex and divide the sample into 1 mL aliquots by dispensing into 2.0 mL conical screw microtubes.

4.2 Cell separation:
Centrifuge the microtubes at 16,000 × *g*, room temperature for 10 minutes. Discard the supernatants in the contaminated waste container. Do not disturb the pellets.

4.3 Cell disruption:
To each pellet, add 2 g of glass beads (use 1 mm glass beads for human sample or 0.1 mm glass beads for animal sample) and 1 mL of extraction buffer. Insert the microtubes in the Mini-beadbeater. Homogenize the samples by beadbeating at maximal speed for 2.5 min in a 4°C room (Note: The 4° C room helps to avoid excessive sample heating from the vigorous mechanical shaking of the bead-beater). Cool the tubes on ice for 1 min and beat for another 2.5 min.

4.4 Centrifuge the tubes at 16,000 × *g*, room temperature for 3 minutes. Transfer the supernatants into 1.5 mL sterile polypropylene microtubes (Note: Centrifugation pellets the glass beads at the bottom of the tubes so the supernatant can be pipetted off).

4.5 To avoid loss of DNA, repeat step 4.4 and 4.5 for each wastewater pellet.

4.6 Genomic DNA purification:

4.6.1 To precipitate the impurities, add ammomium acetate (2 M final concentration) to the supernatant fractions and mix five times by inversion.

4.6.2 Chill the mixture on ice for 5 minutes. Centrifuge the microtubes at 16,000 × *g*, room temperature for 10 min. Transfer the supernatants into 1.5 mL sterile polypropylene microtubes.

4.6.3 Repeat step 4.6.2 to eliminate all the residual impurities in the supernatant fractions.

4.7 Add a volume of absolute isopropanol to the supernatant fractions. Centrifuge the microtubes at 16,000 × *g*, room temperature for 15 minutes. Gently discard the supernatant without disturbing the delicate pellets.

4.8 Rinse the pellets with 70% (v/v) ethanol. Do not disturb the pellets. Centrifuge the microtubes at 16,000 × *g*, room temperature for 5 minutes.

4.9 Dissolve the nucleic acids in 400 μl of 1X TE buffer. Re-precipitate the nucleic acid with 2.5 volumes of 100% ethanol. Centrifuge the microtubes at 16,000 × *g*, room temperature for 15 minutes.

4.10 Wash the pellet with 1 mL of 70% (v/v) ethanol. Centrifuge the microtubes at 16,000 × *g*, room temperature for 15 minutes.

4.11 Dry the nucleic acid under vacuum by using a speedvac.

4.12 Dissolve each nucleic acid pellet in 50 μl of sterile 1X TE buffer. Pool the nucleic acid samples together.

Buffer Preparation

A. Extraction Buffer
50 mL of TrisHCl 1M, pH 8.0 (Fisher biotech No. BP-152-5).
10 mL of EDTA 0.5M (Fisher biotech, No. BP118-500)
150 mL of 20% (w/v) SDS (Laboratoire Mat, No. DB0485)

Bring to 1 L final volume with ddH_20 (double-distilled water). Mix well.
Filter through a 0.45 μm filter.
Store at room temperature.

Immediately before an extraction procedure, add RNaseA to a final concentration of 10 μg/mL in the extraction buffer.

B. RNaseA Solution (1 mg/mL)
Add 10 mg of RNaseA (Sigma, No. R-4875) to 10 mL of sterile water. Mix well and store in aliquots at –20°C.

C. Ammonium Acetate 7.5 M
Dissolve 577.5 g of ammonium acetate in ddH_2O to a final volume of 1 liter. Mix well. Filter through a 0.45 μm filter. Store at room temperature.

APPENDIX E

STANDARD PCR AMPLIFICATION PROCEDURES

1.0 Objectives/Purpose

To amplify specific genes (16S rDNA , *cpn60*, *wecE*) from pure bacterial culture DNA or genomic DNA mixtures.

2.0 Safety and Precautions

To avoid contamination, prepare the PCR mixtures with care in a biological flow hood, and wear gloves when handling the different PCR solutions.

3.0 Materials and Equipment

Genomic DNA
GeneAmp PCR System 9700 (Perkin Elmer)
2 μL, 20 μL, and 200 μL micropipettes
0.2 mL PCR tubes-domed cap
Primers solutions (100 pmol/μL)
2'-Deoxyadenosine 5'-Triphosphate 100 mM solution (dATP) (Amersham Biosciences)
2'-Deoxycytidine 5'-Triphosphate 100 mM solution (dCTP) (Amersham Biosciences)
2'-Deoxyguanosine 5'-Triphosphate 100 mM solution (dGTP) (Amersham Biosciences)
2'-Deoxythymidine 5'-Triphosphate 100 mM solution (dTTP) (Amersham Biosciences)
rTaq DNA polymerase (Amersham Biosciences)
10X PCR Buffer (Amersham Biosciences)

4.0 Procedures

4.1. PCR Amplification of the rDNA 16S

4.1.1. Add the following mix into 0.2 mL PCR tubes:
50 ng of purified genomic DNA
5 μl of 10X PCR buffer
5 μl of a 20 mM dNTP solution
1 μl of a 25 μM forward primers F1 solution
1 μl of a 25 μM reverse primers R2 solution
0.5 μl (2.5 units) of *Taq* DNA polymerase
Bring to 50 μl final volume with sterile ddH_20

4.1.2. Place the PCR tubes containing the mix into the GeneAmp PCR System
4.1.3. Set the following conditions on the PCR machine:

Hot start: 5 min at 94°C
40 cycles of amplification: 1 min at 94°C
1 min at 57°C*
2 min at 72°C
Final extension: 7 min at 72°C

4.1.4. Amplicon verification

Verify amplicon size by 1% agarose gel electrophoresis of 5 µl of the PCR reaction (*125*). The size of the anticipated fragment is 528 bp.

4.2. PCR Amplification of the *cpn60* Gene

4.2.1. Add the following mix into the 0.2 mL PCR tubes:

50 ng of purified genomic DNA
5 µl of 10X PCR buffer
5 µl of a 20 mM dNTP solution
1 µl of a 25 µM forward primers wdF solution
1 µl of a 25 µM reverse primers wdR solution
1 µl of 100 mM $MgCl_2$ (Note: The extra amount of Mg improves the yields with the degenerate primers used in this reaction)
0.5 µl (2.5 units) of *Taq* DNA polymerase
Bring to 50 µl final volume with sterile ddH_20

4.2.2. Place the PCR tubes containing the mix into the GeneAmp PCR System.

4.2.3. Set the following conditions on the machine:

Hot start: 5 min at 94°C
40 cycles of amplification: 1 min at 94°C
1 min at 43°C*
1 min at 72°C
Final extension: 7 min at 72°C

4.2.4. Amplicon verification:

Verify amplicon size by 1% agarose gel electrophoresis of 5 µl of the PCR reaction (*125*). The size of the anticipated fragment is 555 bp.

4.3. PCR Amplification of the *wecE* Gene

4.3.1. Add the following mix into the 0.2 mL PCR tubes:

50 ng of purified genomic
5 µL of 10X PCR buffer
5 µL of a 20 mM dNTP solution
1 µL of a 25 µM forward primers wecE1 solution
1 µL of a 25 µM reverse primers wecE2 solution
0.5 µl (2.5 units) of *Taq* DNA polymerase
Bring to 50 µL final volume with sterile ddH_20.

4.3.2. Place the PCR tubes containing the mix into the GeneAmp PCR System.

4.3.3. Set the following conditions on the machine:

Hot start: 5 min at 94°C
40 cycles of amplification: 1 min at 94°C
30 sec at 55°C*
30 sec at 72°C
Final extension: 7 min at 72°C

4.3.4. Amplicon verification:

Verify amplicon size by 1.5% agarose gel electrophoresis of 5 µL of the PCR reaction (*125*). The size of the anticipated fragment is 188 bp.

4.4. PCR Negative Control

4.4.1. For all PCR reactions, one negative control is used. It consists of the complete PCR mix without DNA.
Add the following mix into the 0.2 mL PCR tubes:
5 μL of 10X PCR buffer
5 μL of a 20 mM dNTP solution
1 μL of a 25 μM forward primers wecE1 solution
1 μL of a 25 μM reverse primers wecE2 solution
0.5 μL (2.5 units) of *Taq* DNA polymerase
Bring to 50 μL final volume with sterile ddH_20.

4.4.2. Place the PCR tubes containing the mix into the GeneAmp PCR System with the other tubes.

4.4.3. Set the specific gene tested PCR conditions on the machine

4.4.4. Amplicon verification: Verify the presence of amplified DNA fragments by 1.5% agarose gel electrophoresis of 5 μL of the PCR reaction (*125*). No amplification must be seen, but if amplification occurs, the rest of the tubes are to be considered as contaminated and the amplifications have to be redone with new batches of of reagents and buffers.

* These specific annealing temperatures were determined using a GeneAmp PCR System 9700 (Perkin Elmer) and might differ between different PCR machines. The temperature should be modified if the overall appearance of the amplicon on the agarose gel is not optimal. An ideal PCR reaction is supposed to generate a single specific fragment. If additional fragments are obtained the annealing temperature should be increased. If no amplified fragments are obtained, the annealing temperature should be decreased.

Preparation and Composition of PCR Solutions

A. dNTP 20 mM solution (1 mL)

200 μL of 2'-Deoxyadenosine 5'-Triphosphate 100 mM solution (dATP) (Amersham)
200 μL of 2'-Deoxycytidine 5'-Triphosphate 100 mM solution (dCTP) (Amersham)
200 μL of 2'-Deoxyguanosine 5'-Triphosphate 100 mM solution (dGTP) (Amersham)
200 μL of 2'-Deoxythymidine 5'-Triphosphate 100 mM solution (dTTP) (Amersham)

Bring to 1 mL final volume with sterile ddH_20. Mix well.
Store in aliquots at –20°C.

B. 100 mM $MgCl_2$

Dissolve 20.33 g magnesium chloride in 800 mL ddH_2O.
Adjust volume to 1 liter with ddH_2O.
Divide in aliquots of 100 mL and sterilize by autoclaving.
Store at room temperature.

C. 10X PCR Buffer from Amersham Biosciences No. 31799

100 mM Tris-HCl, pH 9.0, 15 mM $MgCl_2$, and 500 mM KCl

APPENDIX F

STANDARD LABELING PROCEDURE

1. **Objective/Purpose**
   To label amplicons for microarray hybridization.

2. **Safety**
   Wear gloves when handling the labeling solutions to avoid contamination.
   Keep the Mirus label IT Cy5 and the labeled amplicons in the dark.

3. **Materials and Equipment**
   Speedvac (Savant, Model No. SVC200H)
   Centrifuge 16,000 × *g* (Eppendorf ; Angled-centrifuge 5415)
   37°C incubator
   Amplicons purified by Qiagen PCR purification kit
   Mirus' Cy5 *Label* IT Nucleic Acid Labeling kit
   1.5 mL sterile polypropylene microtubes (Eppendorf)
   2 µL, 20µL, and 200 µL micropipettes
   Sterile deionized water
   Qiagen PCR purification kit

4. **Procedure**

4.1 Labeling

4.1.1 Dry the necessary quantity (2 µg) of purified amplicons in a 1.5 mL sterile polypropylene microtube, by using a speedvac.

4.1.2 Resuspend the dried amplicons in 24.0 µL of sterile ddH_2O, 4.0 µl of 10X buffer A (Mirus kit) and 2 µl of 2X Mirus Cy5 *Label* IT Nucleic Acid Labeling Kit.

4.1.3 Do a quick spin (i.e., 2 sec at 13,200 × *g*) using the centrifuge and incubate at 37°C for 3 hours.

4.1.4 As the labeling reaction was performed in a small volume (30 µL), a quick spin was done after 30 min of incubation to minimize evaporation and keep the concentration of the reaction components at the appropriate levels.

4.2 Purification of Labeled Amplicons

4.2.1 Bring the volume of the labeling reaction to 100 µL with 65.0 µL of sterile water.

4.2.2 Add 5.0 µl of 3 M sodium acetate pH 5.2 to bring the pH of the labeling solution to 8.0 (the optimum pH needed for the Qiagen PCR purification columns).

4.2.3 Add four volumes of PB buffer (Qiagen kit)

4.2.4 The sample was vortexed for five seconds to ensure a good mix, and transferred to a Qiagen column. The column was centrifuged 15 seconds at 15,500 × *g* and the eluant was discarded.

4.2.5 Do three washes with 500 µL of PE Buffer (Qiagen kit) with a centrifugation step at 15,500 × *g* for 15 sec between washes. After the final wash, an additional centrifugation was carried out to ensure that no residual ethanol from the PE buffer remains on the column.

4.2.6 The column is then placed in a new collection tube (1.5 mL Eppendorf tube) and two serial elutions of 30 µL with 1/10 diluted EB buffer (Qiagen kit) were carried out to remove the purified DNA from the column. The elution buffer was added directly to the column surface and allowed to equilibrate for 1 min at room temperature prior to centrifuging 1 minute at 15,500 × *g*.

4.2.7 Keep the labeled amplicons at -20°C.

APPENDIX G

STANDARD MICROARRAY HYBRIDIZATION PROCEDURE

1. **Objective/Purpose**
   To hybridize labeled amplicons on the microarray in order to detect several microorganisms.

2. **Safety**
   Wear gloves when handling the hybridization solutions to avoid contamination.

3. **Materials and Equipment**
   Speedvac (Savant, Model No. SVC200H)
   Centrifuge 16,000 × *g* (Eppendorf ; Angled-centrifuge 5415)
   Rotary hybridization oven (Boekel Scientific, Boekel Bambino)
   95°C water bath
   37°C water bath
   Microarrays
   Labeled amplicons
   22×22 mm coverslips (Sigma)
   Hybridization chamber (Corning)
   50 mL sterile polypropylene conical tubes with cap for centrifugation
   1.5 mL sterile polypropylene microtubes (Eppendorf)
   2 µl, 20µl and 200 µl micropipettes
   Mirus Cy5 *Label* IT Nucleic Acid Labeling kit
   DIG Easy Hyb buffer (Roche Laboratories)
   Salmon sperm DNA solution 10 mg/mL (Invitrogen)
   0.1X SSC buffer
   0.1% (w/v) SDS buffer
   0.1X SSC / 0.1% (w/v) SDS buffer

4. **Procedure**

   4.1 Prehybridation

   4.1.1 Mix in a 1.5 mL sterile polypropylene microtube:
   14 µl of DIG Easy Hyb buffer prewarmed to 37°C.
   1 µl of salmon sperm DNA (10 mg/mL) denaturated by incubation at 95°C for 2 minutes.

   4.1.2 Transfer the mix on the surface of a 22x22 mm coverslip.

   4.1.3 Apply the microarray on the coverslip. Avoid the presence of air bubbles.

4.1.4 Transfer the microarray-coverslip complex into the hybridization chamber and incubate in a 37°C water bath for 1 hr.

4.2 Hybridation

4.2.1 Dry the necessary quantity of fluorecently-labeled amplicons, in a 1.5 mL sterile polypropylene microtube, by using a speedvac.

4.2.2 Resuspend the dried amplicons in 10.5 µL of pre-warmed DIG Easy Hyb buffer and 1 µl of salmon sperm DNA.

4.2.3 Chemically denature the DNA by the addition of 1.5 µL of D1 solution (provided in the Mirus Cy5 *Label* IT Nucleic Acid Labeling kit) and incubate at room temperature for 5 min.

4.2.4 Meanwhile, take out the hybridization chamber (step 4.1.4) from the 37°C bath. Remove the microarray-coverslip complex and remove the coverslip of the microarray by dipping the slide in 0.1X SSC (pre-warmed at 37°C). Dry the slide by centrifugation at room temperature in 50 mL conical tubes for 5 min at 150 X *g*.

4.2.5 Cool the tube containing the chemically denatured DNA on an ice-ethanol mix for 30 sec.

4.2.6 To stop the denaturation reaction, add 1.5 µl of N1 solution (provided in the Mirus Cy5 *Label* IT Nucleic Acid Labeling kit) to the cold denatured DNA mix and incubate at room temperature for 5 min.

4.2.7 Transfer the mix on a 22x22 mm coverslip.

4.2.8 Apply the microarray over the coverslip. Avoid generating air bubbles.

4.2.9 Transfer the microarray-coverslip assembly into the hybridization chamber and incubate in a 37°C water bath for 4 hr.

4.3 **Washing of the Microarray**

4.3.1 After hybridization, remove the coverslip on the microarray by dipping the slide in 0.1X SSC (prewarmed to 37°C).

4.3.2 Transfer the microarray slide into 50 mL conical tubes filled with 0.1X SSC/0.1% (w/v) SDS solution (pre-warmed at 37°C). Place the tubes into a hybridization oven adjusted to 25°C at minimal rotation speed for 10 min.

4.3.3 Repeat step 4.3.2 two times.

4.3.4 Transfer the microarray slide into a 50 mL conical tube filled with 0.1X SSC (prewarmed to 37°C). Place the tube into a hybridization oven adjusted to 25°C at minimal rotation speed for 10 minutes.

4.3.5 Dry the microarray slide by centrifugation (150 x *g*, 5 min, room temperature).

Preparation and Composition of the Hybridization Solutions

A. 20X SSC Buffer
175.32 g of sodium chloride (3M) (Laboratoire MAT, No. SR-0091)
88.23 g of sodium citrate in ddH_2O (0.3M) (Laboratoire MAT, No. SR0194)

Bring to 1 L final volume with ddH_20. Mix well.
Filter through a 0.45 µm filter.
Store at room temperature.

B. 10% (w/v) SDS Buffer
Note: Wear a mask when weighing SDS and clean the weighing area and balance after use.

Dissolve 100 g sodium dodecyl sulfate crystals (SDS) in 900 mL ddH_2O
Heat to 68°C to solubilize the crystals.
Adjust pH to 7.2 with HCl (~ 50 µl)
Adjust volume to 1 L with ddH_2O
Store at room temperature.

REFERENCES

1) Ashbolt, N.J., Grabow, W.O.K., Snozzi, M. In *Water Quality: Guidelines, Standards and Health (World Health Organization)*; Fewtrell, L., Bartram, J., Eds.; IWA Publishing: London, UK, **2001**, pp 289-316.

2) Toranzos, G.A., McFeters, G.A. In *Manual of Environmental Microbiology*; Hurst, C.J., Knudsen, G.R., McInerney, M.J., Stetzenbach, L.D.,Walter, M.V., Eds.; ASM Press: Washington D.C., **1997**, pp 184-194.

3) Venkateswaran, K., Kamijoh, Y., Ohashi, E., Nakanishi, H. A simple filtration technique to detect enterohemorrhagic *Escherichia coli* O157:H7 and its toxins in beef by multiplex PCR. *Appl Environ Microbiol* **1997**, *63*, 4127-4131.

4) Harwood, V.J., Whitlock, J., Withington, V. Classification of antibiotic resistance patterns of indicator bacteria by discriminant analysis: use in predicting the source of fecal contamination in subtropical waters. *Appl Environ Microbiol* **2000**, *66*, 3698-3704.

5) McFeters, G.A., Bissonnette, G.K., Jezeski, J.J., Thomson, C.A., Stuart, D.G. Comparative survival of indicator bacteria and enteric pathogens in well water. *Appl Microbiol* **1974**, *27*, 823-829.

6) Ferguson, C.M., Coote, B.G., Ashbolt, N.J., Stevenson, I.M. Relationships between indicators, pathogens and water quality in an estuarine system. *Water Res* **1996**, *30*, 2045-2054.

7) Griffin, D.W., Gibson, C.J., 3rd, Lipp, E.K., Riley, K., Paul, J.H., 3rd, Rose, J.B. Detection of viral pathogens by reverse transcriptase PCR and of microbial indicators by standard methods in the canals of the Florida Keys. *Appl Environ Microbiol* **1999**, *65*, 4118-4125.

8) Morinigo, M.A., Munoz, M.A., Cornax, R., Martinez-Manzanares, E., Borrego, J.J. Presence of indicators and *Salmonella* in natural waters affected by outfall wastewater discharges. *Water Sci Technol* **1992**, *25*, 1-8.

9) Payment, P., Berte, A., Prévost, M., Ménard, B., Barbeau, B. Occurrence of pathogenic microorganisms in the Saint Lawrence river (Canada) and comparison of health risks for populations using it as their source of drinking water. *Can J Microbiol* **2000**, *46*, 565-576.

10) Davies, C.M., Long, J.A., Donald, M., Ashbolt, N.J. Survival of fecal microorganisms in marine and freshwater sediments. *Appl Environ Microbiol* **1995**, *61*, 1888-1896.

11) Sinton, L.W., Finlay, R.K., Hannah, D.H. Distinguishing human from animal faecal contamination in water: a review. *N Z J Mar Freshwater Res* **1998**, *32*, 323-348.

12) Rutkowski, A.A., Sjogren, R.E. Streptococcal population profiles as indicators of water quality. *Water Air Soil Poll* **1987**, *34*, 273-284.

13) Sargeant, D. *Fecal contamination source identification methods in surface water,* Washington State Department of Ecology, **1999**.

14) Geldreich, E., Kenner, B. Concepts of fecal streptococci in stream pollution. *J Water Pollut Control Fed* **1969**, *41*, 336-352.

15) Resnick, I.G., Levin, M.A. Assessment of bifidobacteria as indicators of human fecal pollution. *Appl Environ Microbiol* **1981**, *42*, 433-438.

16) Gyllenberg, H., Niemela, S., Sormunen, T. Survival of bifidobacteria in water as compared with that of coliforms bacteria and enterococci. *Appl Microbiol* **1960**, *40*, 697-700.

17) Gavini, F., Pourcher, A.M., Neut, C., Monget, D., Romond, C., Oger, C., Izard, D. Phenotypic differentiation of bifidobacteria of human and animal origins. *Int J Syst Bacteriol* **1991**, *41*, 548-557.

18) Mara, D.D., Oragui, J.I. Sorbitol-fermenting bifidobacteria as specific indicators of human faecal pollution. *J Appl Bacteriol* **1983**, *55*, 349-357.

19) Bradley, G., Carter, J., Gaudie, D., King, C. Distribution of the human faecal bacterium *Bacteroides fragilis,* its bacteriophages and their relationship to current sewage pollution indicators in bathing water. *J Appl Microbiol* **1999**, *85*, 90S-100S.

20) Kreader, C.A. Design and evaluation of *Bacteroides* DNA probes for the specific detection of human fecal pollution. *Appl Environ Microbiol* **1995**, *61*, 1171-1179.

21) Fiksdal, L., Maki, J.S., LaCroix, S.J., Staley, J.T. Survival and detection of *Bacteroides* spp., prospective indicator bacteria. *Appl Environ Microbiol* **1985**, *49*, 148-150.

22) Jofre, J., Bosch, A., Lucena, F., Girones, R., Tartera, C. Evaluation of *Bacteroides fragilis* as indicators of the cirological quality of water. *Water Sci Technol* **1986**, *18*, 167-173.

23) Puig, A., Queralt, N., Jofre, J., Araujo, R. Diversity of *Bacteroides fragilis* strains in their capacity to recover phages from human and animal wastes and from fecally polluted wastewater. *Appl Environ Microbiol* **1999**, *65*, 1772-1776.

24) Furuse, K. In *Phage Ecology*; Goyal, S.M., Gerba, C.P.,Bitton, G., Eds.; Wiley Interscience, **1987**, pp 87-124.

25) Havelaar, A.H., Pot-Hogeboom, W.M., Furuse, K., Pot, R., Hormann, M.P. F-specific RNA bacteriophages and sensitive host strains in faeces and wastewater of human and animal origin. *J Appl Bacteriol* **1990**, *69*, 30-37.

26) Beekwilder, J., Nieuwenhuizen, R., Havelaar, A.H., van Duin, J. An oligonucleotide hybridization assay for the identification and enumeration of F-specific RNA phages in surface water. *J Appl Bacteriol* **1996**, *80*, 179-186.

27) Hsu, F.C., Shieh, Y.S., van Duin, J., Beekwilder, M.J., Sobsey, M.D. Genotyping male-specific RNA coliphages by hybridization with oligonucleotide probes. *Appl Environ Microbiol* **1995**, *61*, 3960-3966.

28) Schaper, M., Jofre, J. Comparison of methods for detecting genotypes of F-specific RNA bacteriophages and fingerprinting the origin of faecal pollution in water samples. *J Virol Methods* **2000**, *89*, 1-10.

29) Jagals, P., Grabow, W.O.K., deVillers, J.C. Evaluation of indicators for assessment of human and animal faecal pollution of surface run-off. *Water Sci Technol* **1995**, *31*, 235-241.

30) Chauret, C., Springthorpe, S., Sattar, S. Fate of *Cryptosporidium* oocysts, *Giardia* cysts, and microbial indicators during wastewater treatment and anaerobic sludge digestion. *Can J Microbiol* **1999**, *45*, 257-262.

31) Lemarchand, K., Lebaron, P. Occurrence of *Salmonella* spp and *Cryptosporidium* spp in a French coastal watershed: relationship with fecal indicators. *FEMS Microbiol Lett* **2003**, *218*, 203-209.

32) Morinigo, M.C., R. Munoz, M.A., Romero, P.; Borrego, J.J. Relationships between *Salmonella* spp and indicator microorganisms in polluted natural waters. *Water Res* **1990**, *24*, 117-120.

33) Polo, F., Figueras, M.J., Inza, I., Sala, J., Fleisher, J.M., Guarro, J. Relationship between presence of *Salmonella* and indicators of faecal pollution in aquatic habitats. *FEMS Microbiol Lett* **1998**, *160*, 253-256.

34) Grabow, W. Waterborne diseases: update on water quality assessment and control. *Water SA* **1996**, *22*, 193-202.

35) Leclerc, H., Schwartzbrod, L., Dei-Cas, E. Microbial agents associated with waterborne diseases. *Crit Rev Microbiol* **2002**, *28*, 371-409.

36) Moe, C.L. In *Manual of Environmental Microbiology*; Hurst, C.J., Knudsen, G.R., McInerney, M.J., Stetzenbach, L.D.,Walter, M.V., Eds.; ASM Press: Washington, D.C., **1997**, pp 184-194.

37) Theron, J., Cloete, T.E. Emerging waterborne infections: contributing factors, agents, and detection tools. *Crit Rev Microbiol* **2002**, *28*, 1-26.

38) Lemarchand, K., Parthuisot, N., Catala, P., Lebaron, P. Comparative assessment of epifluorescence microscopy, flow cytometry and solid-phase cytometry used in the enumeration of specific bacteria in water. *Aquat Microb Ecol* **2001**, *25*, 301-309.

39) Rompre, A., Servais, P., Baudart, J., de-Roubin, M.R., Laurent, P. Detection and enumeration of coliforms in drinking water: current methods and emerging approaches. *J Microbiol Methods* **2002**, *49*, 31-54.

40) Josephson, K.L., Gerba, C.P., Pepper, I.L. Polymerase chain reaction detection of nonviable bacterial pathogens. *Appl Environ Microbiol* **1993**, *59*, 3513-3515.

41) Masters, C.I., Shallcross, J.A., Mackey, B.M. Effect of stress treatments on the detection of *Listeria monocytogenes* and enterotoxigenic *Escherichia coli* by the polymerase chain reaction. *J Appl Bacteriol* **1994**, *77*, 73-79.

42) Brousseau, R., Hill, J.E., Prefontaine, G., Goh, S.H., Harel, J., Hemmingsen, S.M. *Streptococcus suis* serotypes characterized by analysis of chaperonin 60 gene sequences. *Appl Environ Microbiol* **2001**, *67*, 4828-4833.

43) Hill, J.E., Seipp, R.P., Betts, M., Hawkins, L., Van Kessel, A.G., Crosby, W.L., Hemmingsen, S.M. Extensive profiling of a complex microbial community by high-throughput sequencing. *Appl Environ Microbiol* **2002**, *68*, 3055-3066.

44) Chizhikov, V., Rasooly, A., Chumakov, K., Levy, D.D. Microarray analysis of microbial virulence factors. *Appl Environ Microbiol* **2001**, *67*, 3258-3263.

45) Heyndrickx, M., Rijpens, N., Herman, L. Molecular detection and typing of foodborne bacterial pathogens: a review. *Appl Microbiol* **2001**, *VOL 2*, 193-238.

46) Yaron, S., Matthews, K.R. A reverse transcriptase-polymerase chain reaction assay for detection of viable *Escherichia coli* O157:H7: investigation of specific target genes. *J Appl Microbiol* **2002**, *92*, 633-640.

47) Kuhnert, P., Boerlin, P., Frey, J. Target genes for virulence assessment of *Escherichia coli* isolates from water, food and the environment. *FEMS Microbiol Rev* **2000**, *24*, 107-117.

48) Campbell, G.R., Prosser, J., Glover, A., Killham, K. Detection of *Escherichia coli* O157:H7 in soil and water using multiplex PCR. *J Appl Microbiol* **2001**, *91*, 1004-1010.

49) Oberst, R.D., Hays, M.P., Bohra, L.K., Phebus, R.K., Yamashiro, C.T., Paszko-Kolva, C., Flood, S.J.A., Sargeant, J.M., Gillepsie, J.R. PCR-based DNA amplification and presumptive detection of *Escherichia coli* O157:H7 with an internal fluorogenic probe and the 5' Nuclease (TaqMan) assay. *Appl Environ Microbiol* **1998**, *64*, 3389-3396.

50) Regnault, B., Martin-Delautre, S., Lejay-Collin, M., Lefevre, M., Grimont, P.A. Oligonucleotide probe for the visualization of *Escherichia coli/Escherichia fergusonii* cells by *in situ* hybridization: specificity and potential applications. *Res Microbiol* **2000**, *151*, 521-533.

51) Small, J., Call, D.R., Brockman, F.J., Straub, T.M., Chandler, D.P. Direct detection of 16S rRNA in soil extracts by using oligonucleotide microarrays. *Appl Environ Microbiol* **2001**, *67*, 4708-4716.

52) Cangelosi, G.A., Brabant, W.H. Depletion of pre-16S rRNA in starved *Escherichia coli* cells. *J Bacteriol* **1997**, *179*, 4457-4463.

53) McKillip, J.L., Jaykus, L.A., Drake, M. rRNA stability in heat-killed and UV-irradiated enterotoxigenic *Staphylococcus aureus* and *Escherichia coli* O157:H7. *Appl Environ Microbiol* **1998**, *64*, 4264-4268.

54) Farnleitner, A.H., Kreuzinger, N., Kavka, G.G., Grillenberger, S., Rath, J., Mach, R.L. Simultaneous detection and differentiation of *Escherichia coli* populations from environmental freshwaters by means of sequence variations in a fragment of the beta-D-glucuronidase gene. *Appl Environ Microbiol* **2000**, *66*, 1340-1346.

55) Leung, K., Topp, E. Bacterial community dynamics in liquid swine manure during storage: molecular analysis using DGGE/PCR of 16S rDNA. *FEMS Microbiol Ecol* **2001**, *38*, 169-177.

56) Becker, P.M., Wand, H., Martius, G.G., Weissbrodt, E., Stottmeister, U. Functional and structural successions in arbitrary samples of heterotrophic bacteria during aerobic treatments of lignite-carbonization wastewater in in situ enclosures. *Can J Microbiol* **1998**, *44*, 211-220.

57) Zimmer, M., Barnhart, H., Idris, U., Lee, M.D. Detection of *Campylobacter jejuni* strains in the water lines of a commercial broiler house and their relationship to the strains that colonized the chickens. *Avian Dis* **2003**, *47*, 101-107.

58) Bernhard, A.E., Field, K.G. A PCR assay to discriminate human and ruminant feces on the basis of host differences in *Bacteroides-Prevotella* genes encoding 16S rRNA. *Appl Environ Microbiol* **2001**, *66*, 4571-4574.

59) Reynolds, K.A., Gerba, C.P., Abbaszadegan, M., Pepper, L.L. ICC/PCR detection of enteroviruses and hepatitis A virus in environmental samples. *Can J Microbiol* **2001**, *47*, 153-157.

60) Bej, A.K., McCarty, S.C., Atlas, R.M. Detection of coliform bacteria and *Escherichia coli* by multiplex polymerase chain reaction: comparison with defined substrate and plating mathods for water quality monitoring. *Appl Environ Microbiol* **1991**, *57*, 2429-2432.

61) Call, D.R., Brockman, F.J., Chandler, D.P. Detecting and genotyping Escherichia coli O157 : H7 using multiplexed PCR and nucleic acid microarrays. *International Journal of Food Microbiology* **2001**, *67*, 71-80.

62) Bej, A.K., Steffan, R.J., DiCesare, J., Haff, L., Atlas, R.M. Detection of coliform bacteria in water by polymerase chain reaction and gene probes. *Appl Environ Microbiol* **1990**, *56*, 307-314.

63) Dombek, P.E., Johnson, L.K., Zimmerley, S.T., Sadowsky, M.J. Use of repetitive DNA sequences and the PCR to differentiate *Escherichia coli* isolates from human and animal sources. *Appl Environ Microbiol* **2000**, *66*, 2572-2577.

64) Tamanai-Shacoori, Z., Jolivet-Gougeon, A., Pommepuy, M., Cormier, M., Colwell, R.R. Detection of enterotoxigenic *Escherichia coli* in water by polymerase chain reaction amplification and hybridization. *Can J Microbiol* **1994**, *40*, 243-249.

65) Tsai, Y.-L., Palmer, C.J., Sangermano, L.R. Detection of *Escherichia coli* in sewage and sludge by polymerase chain reaction. *Appl Environ Microbiol* **1993**, *59*, 353-357.

66) Tsen, H.Y., Lin, C.K., Chi, W.R. Development and use of 16S rRNA gene targeted PCR primers for the identification of *Escherichia coli* cells in water. *J Appl Microbiol* **1998**, *85*, 554-560.

67) Orlandi, P.A., Lampel, K.A. Extraction-free, filter-based template preparation for rapid and sensitive PCR detection of pathogenic parasitic protozoa. *J Clin Microbiol* **2000**, *38*, 2271-2277.

68) Ramsay, G. DNA chips: State-of-the-art. *Nat Biotechnol* **1998**, *16*, 40-44.

69) Letowski, J., Brousseau, R., Masson, L. DNA microarray applications in environmental microbiology. *Anal Lett* **2003**, *36*, 3165-3184.

70) Ye, R.W., Wang, T., Bedzyk, L., Croker, K.M. Applications of DNA microarrays in microbial systems. *J Microbiol Methods* **2001**, *47*, 257-272.

71) Call, D.R., Borucki, M.K., Loge, F.J. Detection of bacterial pathogens in environmental samples using DNA microarrays. *J Microbiol Methods* **2003**, *53*, 235-243.

72) Zhou, J., Thompson, D.K. Challenges in applying microarrays to environmental studies. *Curr Opin Biotechnol* **2002**, *13*, 204-207.

73) Greer, C.W., Whyte, L.G., Lawrence, J.R., Masson, L., Brousseau, R. Genomics technologies for environmental science. *Environ Sci Technol* **2001**, *35*, 364A-370A.

74) Somerville, C.C., Knight, I.T., Straube, W.L., Colwell, R.R. Simple, rapid method for direct isolation of nucleic acids from aquatic environments. *Appl Environ Microbiol* **1989**, *55*, 548-554.

75) Bourrain, M., Achouak, W., Urbain, V., Heulin, T. DNA extraction from activated sludges. *Curr Microbiol* **1999**, *38*, 315-319.

76) Miller, D.N., Bryant, J.E., Madsen, E.L., Ghiorse, W.C. Evaluation and optimization of DNA extraction and purification procedures for soil and sediment samples. *Appl Environ Microbiol* **1999**, *65*, 4715-4724.

77) Kelly, J.J., Chernov, B.K., Tovstanovsky, I., Mirzabekov, A.D., Bavykin, S.G. Radical-generating coordination complexes as tools for rapid and effective fragmentation and fluorescent labeling of nucleic acids for microchip hybridization. *Anal Biochem* **2002**, *311*, 103-118.

78) Wilson, W.J., Strout, C.L., DeSantis, T.Z., Stilwell, J.L., Carrano, A.V., Andersen, G.L. Sequence-specific identification of 18 pathogenic microorganisms using microarray technology. *Mol Cell Probes* **2002**, *16*, 119-127.

79) Bekal, S., Brousseau, R., Masson, L., Prefontaine, G., Fairbrother, J., Harel, J. Rapid identification of *Escherichia coli* pathotypes by virulence gene detection with DNA microarrays. *J Clin Microbiol* **2003**, *41*, 2113-2125.

80) Kricka, L.J. Stains, labels and detection strategies for nucleic acids assays. *Ann Clin Biochem* **2002**, *39 Part 2*, 114-129.

81) Taroncher Oldenburg, G., Griner, E.M., Francis, C.A., Ward, B.B. Oligonucleotide microarray for the study of functional gene diversity in the nitrogen cycle in the environment. *Appl Environ Microbiol* **2003**, *69*, 1159-1171.

82) Wu, L., Thompson, D.K., Li, G., hurt, R.A., Tiedje, J.M., Zhou, J. Development and evaluation of functionnal gene arrays for detection of selected genes in the environment. *Appl Environ Microbiol* **2001**, *67*, 5780-5790.

83) Dennis, P., Edwards, E.A., Liss, S.N., Fulthorpe, R. Monitoring gene expression in mixed microbial communities by using DNA microarrays. *Appl Environ Microbiol* **2003**, *69*, 769-778.

84) MathieuDaude, F., Welsh, J., Vogt, T., McClelland, M. DNA rehybridization during PCR: The 'C(O)t effect' and its consequences. *Nucleic Acids Res* **1996**, *24*, 2080-2086.

85) Polz, M.F., Cavanaugh, C.M. Bias in template-to-product ratios in multitemplate PCR. *Appl Environ Microbiol* **1998**, *64*, 3724-3730.

86) van Ijperen, C., Kuhnert, P., Frey, J., Clewley, J.P. Virulence typing of Escherichia coli using microarrays. *Mol Cell Probes* **2002**, *16*, 371-378.

87) Gingeras, T.R., Ghandour, G., Wang, E.G., Berno, A., Small, P.M., Drobniewski, F., Alland, D., Desmond, E., Holodniy, M., Drenkow, J. Simultaneous genotyping and species identification using hybridization pattern recognition analysis of generic *Mycobacterium* DNA arrays. *Genome Res* **1998**, *8*, 435-448.

88) Stine, O.C., Carnahan, A., Singh, R., Powell, J., Furuno, J.P., Dorsey, A., Silbergeld, E., Williams, H.N., Morris, J.G. Characterization of microbial communities from coastal waters using microarrays. *Environ Monit Assess* **2003**, *81*, 327-336.

89) Wang, R.F., Kim, S.J., Robertson, L.H., Cerniglia, C.E. Development of a membrane-array method for the detection of human intestinal bacteria in fecal samples. *Mol Cell Probes* **2002**, *16*, 341-350.

90) Straub, T.M., Daly, D.S., Wunshel, S., Rochelle, P.A., DeLeon, R., Chandler, D.P. Genotyping *Cryptosporidium parvum* with an *hsp70* single-nucleotide polymorphism microarray. *Appl Environ Microbiol* **2002**, *68*, 1817-1826.

91) Jones, G.F., Ward, G.E., Murtaugh, M.P., Lin, G., Gebhart, C.J. Enhanced detection of intracellular organism of swine proliferative enteritis, ileal symbiont intracellularis, in feces by polymerase chain reaction. *J Clin Microbiol* **1993**, *31*, 2611-2615.

92) Popovic, T., Bopp, C.A., Olsvik, Ø., Kiehlbauch, J.A. *Diagnostic Molecular Microbiology*; ASM: Washington, D.C., **1993**.

93) Bourrain, M., Achouak, W., Urbain, V., Heulin, T. DNA extraction from activated sludges. *Curr Microbiol* **1999**, *38*, 315-319.

94) Yu, Z., Mohn, W.W. Killing two birds with one stone: simultaneous extraction of DNA and RNA from activated sludge biomass. *Can J Microbiol* **1999**, *45*, 269-272.

95) Guillaume, G., Verbrugge, D., Chasseur-Libotte, M., Moens, W., Collard, J. PCR typing of tetracycline resistance determinants (Tet A-E) in *Salmonella enterica* serotype Hadar and in the microbial community of activated sludges from hospital and urban wastewater treatment facilities in Belgium. *FEMS Microbiol Ecol* **2000**, *32*, 77-85.

96) Yeates, C., Gillings, M.R., Davison, A.D., Altavilla, N., Veal, D.A. Methods for microbial DNA extraction from soil for PCR amplification. *Biological Procedures Online (*www.biologicalprocedures.com*)* **1998**.

97) Cebula, T.A., Payne, W.L., Feng, P. Simultaneous identification of strains of *Escherichia coli* serotype O157:H7 and their Shiga-like toxin type by mismatch amplification mutation assay-multiplex PCR. *J Clin Microbiol* **1995**, *33*, 248-250.

98) Dorsch, M., Stachebrandt, E. Some modifications in the procedure of direct sequencing of PCR amplified 16S rDNA. *J Microbiol Meth* **1992**, *16*, 271-279.

99) Chizhikov, V., Rasooly, A., Chumakov, K., Levy, D.D. Microarray analysis of microbial virulence factors. *Appl Environ Microbiol* **2001**, *67*, 3258-3263.

100) Wang, X., Seed, B. Selection of oligonucleotide probes for protein coding sequences. *Bioinformatics* **2003**, *19*, 796-802.

101) Bayardelle, P., Zafarullah, M. Development of oligonucleotide primers for the specific PCR-based detection of the most frequent Enterobacteriaceae species DNA using wec gene templates. *Can J Microbiol* **2002**, *48*, 113-122.

102) Finegold, S.M., Sutter, V.L., Mathisen, G.E. In *Human intestinal microflora in health and disease*; Hentges, D.J., Ed.: New york, **1983**; Vol. 1, p 568.

103) Moore, W.E., Moore, L.V., Cato, E.P., Wilkins, T.D., Kornegay, E.T. Effect of high-fiber and high-oil diets on the fecal flora of swine. *Appl Environ Microbiol* **1987**, *53*, 1638-1644.

104) Salanitro, J.P., Blake, I.G., Muirehead, P.A., Maglio, M., Goodman, J.R. Bacteria isolated from the duodenum, ileum, and cecum of young chicks. *Appl Environ Microbiol* **1978**, *35*, 782-790.

105) Leser, T.D., Amenuvor, J.Z., Jensen, T.K., Lindecrona, R.H., Boye, M., Moller, K. Culture-independent analysis of gut bacteria: the pig gastrointestinal tract microbiota revisited. *Appl Environ Microbiol* **2002**, *68*, 673-690.

106) Krause, D.O., Russell, J.B. How many ruminal bacteria are there? *J Dairy Sci* **1996**, *79*, 1467-1475.

107) Robinson, I.M., Allison, M.J., Bucklin, J.A. Characterization of the cecal bacteria of normal pigs. *Appl Environ Microbiol* **1981**, *41*, 950-955.

108) Devriese, L.A., Hommez, J., Wijfels, R., Haesebrouck, F. Composition of the enterococcal and streptococcal intestinal flora of poultry. *J Appl Bacteriol* **1991**, *71*, 46-50.

109) Lu, J., Sanchez, S., Hofacre, C., Maurer, J.J., Harmon, B.G., Lee, M.D. Evaluation of Broiler Litter with Reference to the Microbial Composition as Assessed by Using 16S rRNA and Functional Gene Markers. *Appl Environ Microbiol* **2003**, *69*, 901-908.

110) Carman, R.J., Van Tassell, R.L., Wilkins, T.D. The normal intestinal microflora: ecology, variability and stability. *Vet Hum Toxicol* **1993**, *35 Suppl 1*, 11-14.

111) Thompson, J.D., Higgins, D.G., Gibson, T.J. CLUSTAL W: improving the sensitivity of progressive multiple sequence alignment through sequence weighting, position-specific gap penalties and weight matrix choice. *Nucleic Acids Res* **1994**, *22*, 4673-4680.

112) Kreader, C.A. Design and evaluation of *Bacteroides* DNA probes for the specific detection of human fecal pollution. *Appl Environ Microbiol* **1995**, *61*, 1171-1179.

113) Wang, R.F., Cao, W.W., Cerniglia, C.E. PCR detection and quantitation of predominant anaerobic bacteria in human and animal fecal samples. *Appl Environ Microbiol* **1996**, *62*, 1242-1247.

114) Matsuki, T., Watanabe, K., Tanaka, R., Fukuda, M., Oyaizu, H. Distribution of bifidobacterial species in human intestinal microflora examined with 16S rRNA-gene-targeted species-specific primers. *Appl Environ Microbiol* **1999**, *65*, 4506-4512.

115) Lynch, P.A., Gilpin, B.J., Sinton, L.W., Savill, M.G. The detection of *Bifidobacterium adolescentis* by colony hybridization as an indicator of human faecal pollution. *J Appl Microbiol* **2002**, *92*, 526-533.

116) Bernhard, A.E., Field, K.G. A PCR assay to discriminate human and ruminant feces on the basis of host differences in *Bacteroides-Prevotella* genes encoding 16S rRNA. *Appl Environ Microbiol* **2000**, *66*, 4571-4574.

117) Wang, R.F., Cao, W.W., Franklin, W., Campbell, W., Cerniglia, C.E. A 16S rDNA-based PCR method for rapid and specific detection of *Clostridium perfringens* in food. *Mol Cell Probes* **1994**, *8*, 131-137.

118) Loge, F.J., Emerick, R.W., Thompson, D.E., Nelson, D.C., Darby, J.L. Development of a flurescent 16S rRNA oligonucleotide probe specific to the family Enterobacteriaceae. *Water Environ Res* **1999**, *71*, 75-83.

119) Ootsubo, M., Shimizu, T., Tanaka, R., Sawabe, T., Tajima, K., Yoshimizu, M., Ezura, Y., Ezaki, T., Oyaizu, H. Oligonucleotide probe for detecting Enterobacteriaceae by in situ hybridization. *J Appl Microbiol* **2002**, *93*, 60-68.

120) Manero, A., Blanch, A.R. Identification of *Enterococcus* spp. based on specific hybridisation with 16S rDNA probes. *J Microbiol Methods* **2002**, *50*, 115-121.

121) Tsen, H.Y., Lin, C.K., Chi, W.R. Development and use of 16S rRNA gene targeted PCR primers for the identification of *Escherichia coli* cells in water. *J Appl Microbiol* **1998**, *85*, 554-560.

122) Banerjee, A., Yasseri, M., Munson, M. A method for the detection and quantification of bacteria in human carious dentine using fluorescent in situ hybridisation. *J Dent* **2002**, *30*, 359-363.

123) Whitehead, T.R., Cotta, M.A. Development of molecular methods for identification of *Streptococcus bovis* from human and ruminal origins. *FEMS Microbiol Lett* **2000**, *182*, 237-240.

124) Health Canada. *Laboratory Biosafety Guidelines*; 2nd ed.; National Council of Welfare, **1996**.

125) Sambrook, J., Russel, D.W. *Molecular cloning: a laboratory manual*; 3rd ed.; Cold Spring Harbor Laboratory Press, **2001**.

WERF SUBSCRIBERS

WASTEWATER UTILITY

Alabama
Montgomery Water Works & Sanitary Sewer Board

Alaska
Anchorage Water & Wastewater Utility

Arizona
Gila Resources

Glendale, City of, Utilities Department

Mesa, City of

Peoria, City of

Pima County Wastewater Management

Arkansas
Little Rock Wastewater Utility

California
Calaveras County Water District

Central Contra Costa Sanitary District

Contra Costa Water District

Crestline Sanitation District

Delta Diablo Sanitation District

Dublin San Ramon Services District

East Bay Dischargers Authority

East Bay Municipal Utility District

El Dorado Irrigation District

Fairfield-Suisun Sewer District

Fresno Department of Public Utilities

Irvine Ranch Water District

Las Virgenes Municipal Water District

Livermore, City of

Lodi, City of

Los Angeles, City of

Los Angeles County, Sanitation Districts of

Napa Sanitation District

Orange County Sanitation District

Palo Alto, City of

Riverside, City of

Sacramento Regional County Sanitation District

San Diego Metropolitan Wastewater Department, City of

San Francisco, City & County of

San Jose, City of

Santa Barbara, City of

Santa Rosa, City of

South Bayside System Authority

South Coast Water District

South Orange County Wastewater Authority

Sunnyvale, City of

Union Sanitary District

West Valley Sanitation District

Colorado
Boulder, City of

Colorado Springs Utilities

Littleton/Englewood Water Pollution Control Plant

Metro Wastewater Reclamation District, Denver

Connecticut
The Mattabasset District

New Haven, City of, WPCA

District of Columbia
District of Columbia Water & Sewer Authority

Florida
Broward, County of

Fort Lauderdale, City of

Gainesville Regional Utilities

JEA

Miami-Dade Water & Sewer Authority

Orange County Utilities Department

Orlando, City of

Reedy Creek Improvement District

Seminole County Environmental Services

St. Petersburg, City of

Stuart Public Utilities

Tallahassee, City of

Tampa, City of

Toho Water Authority

West Palm Beach, City of

Georgia
Atlanta Department of Watershed Management

Augusta, City of

Clayton County Water Authority

Cobb County Water System

Columbus Water Works

Fulton County

Gwinnett County Department of Public Utilities

Savannah, City of

Hawaii
Honolulu, City and County of

Illinois
American Bottoms Wastewater Treatment Plant

Greater Peoria Sanitary District

Kankakee River Metropolitan Agency

Metropolitan Water Reclamation District of Greater Chicago

North Shore Sanitary District

Wheaton Sanitary District

Iowa
Ames, City of

Cedar Rapids Wastewater Facility

Des Moines Metro Wastewater Reclamation Authority

Iowa City, City of

Kansas
Johnson County Unified Wastewater Districts

Unified Government of Wyandotte County/Kansas City, City of

Kentucky
Louisville & Jefferson County Metropolitan Sewer District

Louisiana
Sewerage & Water Board of New Orleans

Maine
Bangor, City of

Portland Water District

Maryland
Anne Arundel County Bureau of Utility Operations

Howard County Department of Public Works

Washington Suburban Sanitary Commission

Massachusetts
Boston Water & Sewer Commission

Upper Blackstone Water Pollution Abatement District

Michigan
Ann Arbor, City of

Detroit, City of

Holland Board of Public Works

Lansing, City of

Owosso Mid-Shiawassee County WWTP

Saginaw, City of

Wayne County Department of Environment

Wyoming, City of

Minnesota
Rochester, City of

Western Lake Superior Sanitary District

Missouri
Independence, City of

Kansas City Missouri Water Services Department

Little Blue Valley Sewer District

Nebraska
Lincoln Wastewater System

Nevada
Henderson, City of

New Jersey
Bergen County Utilities Authority

Passaic Valley Sewerage Commissioners

New York
New York City Department of Environmental Protection

Rockland County Solid Waste Management Authority/ Sewer District

North Carolina
Charlotte/Mecklenburg Utilities

Durham, City of

Metropolitan Sewerage District of Buncombe County

Orange Water & Sewer Authority

Ohio
Akron, City of

Butler County Department of Environmental Services

Columbus, City of

Metropolitan Sewer District of Greater Cincinnati

Northeast Ohio Regional Sewer District

Summit, County of

Oklahoma
Tulsa, City of

Oregon
Clean Water Services

Eugene/Springfield Water Pollution Control

Water Environment Services

Pennsylvania
Philadelphia, City of

University Area Joint Authority, State College

South Carolina
Charleston Commissioners of Public Works

Mount Pleasant Waterworks & Sewer Commission

Spartanburg Sanitary Sewer District

Tennessee
Cleveland, City of

Knoxville Utilities Board

Murfreesboro Water & Sewer Department

Nashville Metro Water Services

Texas
Austin, City of

Dallas Water Utilities

Denton, City of

El Paso Water Utilities

Fort Worth, City of

Gulf Coast Waste Disposal Authority

Houston, City of

San Antonio Water System

Trinity River Authority

Utah
Salt Lake City Corporation

Virginia
Alexandria Sanitation Authority
Arlington, County of
Fairfax County Virginia
Hampton Roads Sanitation District
Henrico, County of
Hopewell Regional Wastewater Treatment Facility
Loudoun County Sanitation Authority
Lynchburg Regional WWTP
Prince William County Service Authority
Richmond, City of
Rivanna Water & Sewer Authority

Washington
Edmonds, City of
Everett, City of
King County Department of Natural Resources
Seattle Public Utilities
Sunnyside, Port of
Yakima, City of

Wisconsin
Green Bay Metro Sewerage District
Madison Metropolitan Sewerage District
Milwaukee Metropolitan Sewerage District
Racine, City of
Sheboygan Regional Wastewater Treatment
Wausau Water Works

Australia
South Australian Water Corp.
Sydney Water Corp.
Water Corp. of Western Australia

Canada
Greater Vancouver Regional District
Toronto, City of, Ontario
Winnipeg, City of, Manitoba

Mexico
Servicios de Agua y Drenaje de Monterrey, I.P.D.

New Zealand
Watercare Services Limited

Singapore
Singapore Public Utilities Board

United Kingdom
Yorkshire Water Services Limited

STORMWATER UTILITY

California
Los Angeles, City of, Department of Public Works
Monterey, City of
San Francisco, City & County of
Santa Rosa, City of
Sunnyvale, City of

Colorado
Boulder, City of

Georgia
Griffin, City of

Iowa
Cedar Rapids Wastewater Facility
Des Moines Metro Wastewater Reclamation Authority

Kansas
Overland Park, City of

Kentucky
Louisville & Jefferson County Metropolitan Sewer District

Maine
Portland Water District

Minnesota
Western Lake Superior Sanitary District

North Carolina
Charlotte, City of, Stormwater Services

Pennsylvania
Philadelphia, City of

Tennessee
Chattanooga Storm Water Management

Washington
Seattle Public Utilities

STATE

Arkansas Department of Environmental Quality
Fresno Metropolitan Flood Control District, Calif.
Urban Drainage & Flood Control District, Colo.

CORPORATE

ADS Environmental Services
The ADVENT Group Inc.
Alan Plummer & Associates
Alden Research Laboratory
Aqua-Aerobic Systems, Inc.
Aquateam–Norwegian Water Technology Centre A/S
BaySaver Inc.
BioVir Laboratories, Inc.
Black & Veatch
Boyle Engineering Corporation
Brown & Caldwell
Burns & McDonnell
CABE Associates Inc.
The Cadmus Group
Camp Dresser & McKee Inc.
Carollo Engineers Inc.
Carpenter Environmental Associates Inc.
CDS Technologies Inc.
Chemtrac Systems Inc.
CH2M HILL
Damon S. Williams Associates, LLC
David L. Sheridan, P.C.
Dewling Associates, Inc.
Earth Tech Inc.
Eco-Matrix
Ecovation
Environmental Engineers International
EMA Inc.
The Eshelman Company Inc.
Finkbeiner, Pettis, & Strout Inc.
Freese & Nichols, Inc.
ftn Associates Inc.
Fuss & O'Neill, Inc.
Gannett Fleming Inc.
Golder Associates Inc.
Greeley and Hansen LLC
Hazen & Sawyer, P.C.
HDR Engineering Inc.
HNTB Corporation
HydroQual Inc.
Infilco Degremont Inc.
Ingersoll-Rand Energy Systems
Insituform Technologies Inc.
Institute for Environmental Technology & Industry, Korea
Jacobson Helgoth Consultants Inc.
Jason Consultants Inc.
Jordan, Jones, & Goulding Inc.
KCI Technologies Inc.
Kelly & Weaver, P.C.
Kennedy/Jenks Consultants
Komline Sanderson Engineering Corporation
Lawler, Matusky & Skelly Engineers, LLP
Limno-Tech Inc.
Lombardo Associates Inc.
Malcolm Pirnie Inc.
McKim & Creed
MEC Analytical Systems Inc.
Metcalf & Eddy Inc.
MWH
New England Organics
Odor & Corrosion Technology Consultants Inc. (OCTC)
Oswald Green, LLC
PA Government Services Inc.
Parametrix Inc.
Parsons
Post, Buckley, Schuh & Jernigan
R&D Engineering/Conestoga Rover & Associates
The RETEC Group
RMC, Inc. (Raines, Melton & Carella)
R.M. Towill Corporation
Ross & Associates Ltd.
Royce Technologies
SAIC Maritime Technical Group
Short Elliott Hendrickson, Inc.
Stantec Consulting Group, Inc.
Stormwater Management, Inc.
Synagro Technologies, Inc.
Tetra Tech Inc.
Trojan Technologies Inc.
URS Corporation
Veolia Water NATC
Wade-Trim Inc.
Weston Solutions Inc.
Woodard & Curran
WRc/D&B, LLC
WWETCO, LLC
Zoeller Pump Company

INDUSTRY

American Electric Power
ChevronTexaco Energy Research & Technology Company
The Coca-Cola Company
Dow Chemical Company
DuPont Company
Eastman Kodak Company
Eli Lilly & Company
Merck & Company Inc.
ONDEO Services
Procter & Gamble Company
PSEG Services Corp.
RWE Thames Water Plc
Severn Trent Services Inc.
United Water Services LLC

Note: List as of 7/1/04

WERF Product Order Form

s a benefit of joining the Water Environment Research Foundation, subscribers are entitled to receive one complimentary copy of all final ports and other products.Additional copies are available at cost (usually $10).To order your complimentary copy of a report, please write ee" in the unit price column.WERF keeps track of all orders.If the charge differs from what is shown here, we will call to confirm the total efore processing.

ame ____________________ Title ____________________

ganization ____________________

ddress ____________________

ty __________ State __________ Zip Code __________ Country __________

one __________ Fax __________ Email __________

Stock #	Product	Quantity	Unit Price	Total
			Postage & Handling	
			VA Residents Add 4.5% Sales Tax	
			Canadian Residents Add 7% GST	
			TOTAL	

Method of Payment: (All orders must be prepaid.)

- ❑ Check or Money Order Enclosed
- ❑ Visa ❑ Mastercard ❑ American Express

ccount No. ____________________ Exp. Date __________

gnature ____________________

Shipping & Handling:

Amount of Order	United States	Canada & Mexico	All Others
Up to but not more than:	Add:	Add:	Add:
$20.00	$5.00*	$8.00	50% of amount
30.00	5.50	8.00	40% of amount
40.00	6.00	8.00	
50.00	6.50	14.00	
60.00	7.00	14.00	
80.00	8.00	14.00	
100.00	10.00	21.00	
150.00	12.50	28.00	
200.00	15.00	35.00	
More than $200.00	Add 20% of order	Add 20% of order	

*minimum amount for all orders

ote: Please make checks payable to the Water Environment Research Foundation.

To Order (Subscribers Only):

Log on to www.werf.org and click on the "Product Catalog."

Phone: (703) 684-2470
Fax: (703) 299-0742.

WERF
Attn: Subscriber Services
635 Slaters Lane
Alexandria, VA 22314-1177

To Order (Non-Subscribers):

Non-subscribers may be able to order WERF publications either through WEF (www.wef.org) or IWAP (www.iwapublishing.com). Visit WERF's website at www.werf.org for details.

IWA Publishing's authorised EU representative for General Product Safety Regulations is Diane D'Arras, 15 rue Duret, 75116 Paris, France, e-mail: safety@iwap.co.uk.

Printed and bound by CPI Group (UK) Ltd, Croydon, CR0 4YY
12/05/2026
02108876-0014